5th International Symposium on Hydrogen Energy, Renewable Energy and Materials (HEREM 2019)

IOP Conference Series: Materials Science and Engineering
Volume 605

Bangkok, Thailand
13 - 14 June 2019

Editors:

Mohan Lal Kolhe

ISBN: 978-1-5108-9482-2
ISSN: 1757-8981

Printed from e-media with permission by:

Curran Associates, Inc.
57 Morehouse Lane
Red Hook, NY 12571

Some format issues inherent in the e-media version may also appear in this print version.

Institute of Physics
Dirac House, Temple Back
Bristol BS1 6BE UK

Phone: 44 1 17 929 7481
Fax: 44 1 17 920 0979

techtracking@iop.org

Additional copies of this publication are available from:

Curran Associates, Inc.
57 Morehouse Lane
Red Hook, NY 12571 USA
Phone: 845-758-0400
Fax: 845-758-2633
Email: curran@proceedings.com
Web: www.proceedings.com

Table of contents

Volume 605

The 5th International Symposium on Hydrogen Energy, Renewable Energy and Materials

13–14 June 2019, Bangkok, Thailand

Accepted papers received: 1 August 2019
Published online: 30 August 2019

Preface

Papers

Hydrogen Energy

Renewable Energy Technologies and System Engineering

IOP Conf. Series: Materials Science and Engineering **605** (2019) 011001 doi:10.1088/1757-899X/605/1/011001

The 5th International Symposium on Hydrogen Energy, Renewable Energy and Materials (HEREM 2019)

June 13-14, 2019 at Bangkok, Thailand

Editorial

The fifth International Symposium on Hydrogen Energy, Renewable Energy and Materials 2019 (HEREM 2019) has been organized and held on June 13-14, 2019 in Bangkok (Thailand). The HEREM 2019 has been sponsored by the International Energy and Environmental Research Institute and technically assisted by the University of Agder (Norway). The aim of this HEREM is to provide unique platform for researchers, scientists, engineers and professionals from all over the world to present and share their latest research results and new ideas in the fields of Hydrogen Energy, Renewable Energy and Materials for sustainable development of the global community.

The HEREM 2019 has been organized with special plenary speeches on advancement of integration of renewable energy technologies with hydrogen as an energy carrier, solar catalysis for hydrogen power and fuel cell, global perspective on economic feasibility of small scale solar photovoltaic systems, role of catalysis for greenhouse gas conversion into fuels, materials advancement in milling tools.

With successful experience and learning from the previous years' HEREM, it is going to contribute consistently in the scientific domain of hydrogen energy and renewable energy technologies with materials engineering for sustainable development. The proceeding of HEREM 2018 was included in the Elsevier's Energy Procedia (vol. 144, pages 1-194).

IOP Conf. Series: Materials Science and Engineering **605** (2019) 011001 doi:10.1088/1757-899X/605/1/011001

The selected peer-reviewed papers of HEREM 2019 have been edited as conference proceedings and to be published with the IOP Conference Series: Materials Science and Engineering (MSE) (Open Access) (ISSN: 1757-8981) DoI: 10.1088/issn.1757-899X. The HEREM 2019 proceeding engrosses articles in the filed of hydrogen energy, renewable energy technologies and system engineering, energy materials engineering.

As the editor of this special issue of the HEREM 2019 proceedings, I wish to express my sincere gratitude to the plenary speakers, authors, participants, reviewers, program and technical committee members, and organizing members; who have contributed in organization of the conference and publication of this special issue. With high-quality submissions and knowledge disseminations in the HEREM 2019, I am confident that the HEREM will become an important conference in this multi-discipline domains of hydrogen energy with renewable energy technologies and materials engineering.

Professor Mohan Lal Kolhe
Editor of the HEREM 2019 Proceeding
Professor in Smart Grid and Renewable Energy
Faculty of Engineering and Science,
University of Agder,
PO Box 422, NO 4604, Kristiansand, Norway
www.uia.no

IOP Conf. Series: Materials Science and Engineering **605** (2019) 011002 doi:10.1088/1757-899X/605/1/011002

Peer review statement

All papers published in this volume of IOP Conference Series: Materials Science and Engineering have been peer reviewed through processes administered by the proceedings Editors. Reviews were conducted by expert referees to the professional and scientific standards expected of a proceedings journal published by IOP Publishing.

IOP Conf. Series: Materials Science and Engineering **605** (2019) 012001 doi:10.1088/1757-899X/605/1/012001

Research and Analysis on Gas Performance Index of Pine Particle Gasification

Yonghong Niu[1,2], Liqi Zhang[1,3], Yike Li[1]*, Songsheng Wang[1],Huijun Wu[1]

[1]School of Energy and Environment, Inner Mongolia University Science and Technology, Baotou 014010, Inner Mongolia, China
[2]Mining Research Institute, Inner Mongolia University Science and Technology, Baotou 014010, Inner Mongolia, China
[3]Baotou Iron&Steel Vocational Technical College,Baotou 014010,Inner Mongolia,China

1047683721@qq.com

Abstract. The gasification test was carried out with pine particle as raw material, how temperature and steam flow rate affect gas gasification performance index was studied by using self-made gasification test system platform. The results show that the volume fraction of hydrogen increases ,the volume percentage of carbon monoxide decreases and CH_4 content decreases with the increase of temperature in steam flow rate of 0.3kg / h;Its the gas volume fraction of the trend is similar with the steam flow rate of 0.3kg / h,but the total gas energy with the temperature increased first and then reduced and the total gas increased by 28393.39KJ / Kg at 800 °C ; The steam flow from 0.3kg / h to 0.6kg / h, the low calorific value of gas varies from 10 to14MJ / m^3, the increase of the calorific value of gas is 2111.36KJ / m^3.

1. Introduction

As a clean energy , biomass has two characteristics: the first, biomass containing sulfur, nitrogen content is low, Gases generated during the utilization and conversion of biomass have low fouling properties. The second, Biomass energy is formed by photosynthesis of green plants, it can effectively reduce the greenhouse gas CO_2, can improve the ecological environment because of releasing O_2 [1-2]. Biomass gasification converts low-grade biomass into clean gas, and replaces non-renewable fuels for human needs[3].Therefore, biomass gasification technology, especially water vapor gasification technology, has received universal attention[4]. Kong Meng et al.[5] think that biomass utilization is one of the most promising utilization methods in energy conversion, so it has become a hot spot for research. The heart of biomass gasification is the gasifier, which is also the least efficient system in gasification systems[6]. The traditional evaluation of gasification efficiency generally uses carbon conversion rate and cold gas efficiency, this method is theoretically deficient [7]. Dinner I et al.[8]showed that in an adiabatic system (ignoring heat loss), the energy efficiency can reach 100% while the exergy efficiency is only 79%. Therefore, many scholars recommend using exergy performance indicators instead of the traditional energy analysis in order to achieve a more reasonable evaluation. Through the research on the characteristics of high-yield gas produced by gasification of biomass and the analysis of gasification process performance indexes (low calorific value of syngas, exergy efficiency and gas exergy), the quality of gas is improved[9].

2. Method

2.1. Experimental Materials

The gasification test was carried out with waste pine sawdust material of a wood processing plant in Baotou (Figure1(a)) by using self-made high temperature steam gasification of biomass to produce hydrogen gas test system platformFigure1(b). Pine sawdust elemental analysis and industrial analysis as shown in Table 1. Pine shredded pellets were passed through a pellet mill (model K120 manufactured by Xiaoyi Mingyang Machinery Factory, Figure1(b)) before the test, Pellets with a grain size of about 10 mm and 10mm long were selected as the test material (Figure1(c)). Exhaust air from the system through nitrogen injection and check its air tightness before experiment. And then the water and gasifier were heated to a predetermined temperature by the control system and tested. Then different steam flow rates and temperatures were selected as variables to test. At the same time, 160 mg particles were weighed, and the nitrogen purity of 99.99% and 100 ml / min of flow rate was used as a carrier. Thermal gravimetric analysis was carried out by heating to a predetermined temperature of 800°C. at a heating rate of 10K/min. The weight loss was measured at different times. The different stages in the gasification and pyrolysis process are studied.

(a) (b) (c)

Figure 1. Experimental raw materials and equipment.

Table 1. Proximate and ultimate analysis of pine wood shavings.

Elemental Analysis	Element	C_{ad}	H_{ad}	O_{ad}	N_{ad}	S_{ad}
	mass fraction/%	50.54	7.08	41.11	0.15	0.57
Industry Analysis	Compositio n	V^a	FC^b	A^c	Q_{LHV}^d MJ/kg	
	mass fraction/%	82.29	17.16	0.55	16.89	

[a] V—Volatile

[b]FC—Fixed Carbon

[c]A—Ash

[d]Q_{LHV}—Lower Calorific value

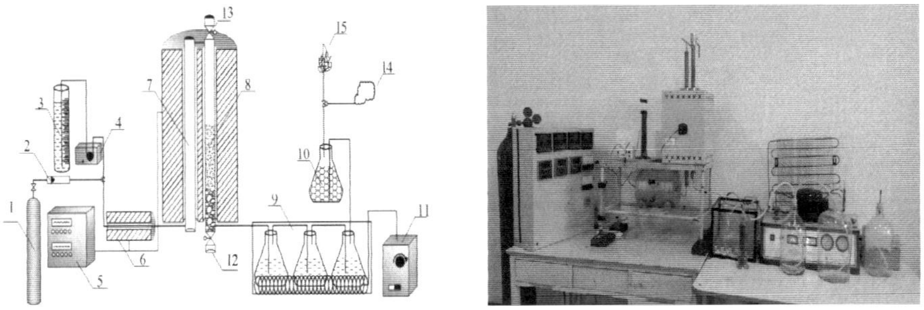

1. nitrogen bottle 2. rotameter 3. water tank 4. kamoeradjustable speed creep micro pump 5. temperature control cabinet 6. steam generator 7. steam heating device 8.biomass gasification reactor 9. tar absorption device 10. drying device 11. refrigeration equipment 12. discharge port 13. packing mouth 14. air bag 15. gas processing plant

Figure 2. Biomass steam gasification experiment platform.

2.2. Determination of gasification components and component content

Gas chromatography (Agilen 7890B) manufactured by Agilent Technologies Inc. was used to detect and analyze pine gas at different flow rates and temperatures, and gas analysis was performed using a thermal conductivity detector (TCD) such as H_2, CO and CH_4.

2.3. Analysis of Gasification Process Performance Indexes

For the gasification system, the input of exergy to the gasifier include biomass exergy flow rate and steam exergy flow rate, the output of exergy are gas exergy flow rate, tar exergy flow rate, and ash exergy flow rate. For gasifiers exergy balance equation is

$$\dot{E}_{steam} + \dot{E}_{biomass} - \dot{E}_{fuel} - \dot{E}_{ash} - \dot{E}_{tar} = \dot{E}_{dest} \quad (1)$$

Where $\dot{E}_{steam}$ is steam exergy flow rate, KW; $\dot{E}_{biomass}$ is biomass exergy flow rate,KW; $\dot{E}_{fuel}$ is gas exergy flow rate,KW; $\dot{E}_{ash}$ is ash exergy flow rate,KW; $\dot{E}_{tar}$ is tar exergy flow rate, KW; $\dot{E}_{dest}$ is gasifier exergy loss.

2.3.1. Gas exergy flow rate . Gas exergy flow rate includes physical exergy flow rate and chemical exergy flow rate, which are calculated as:

$$\dot{E}_{fuel} = \dot{E}_{ph} + \dot{E}_{ch} \quad (2)$$

in this equation, E_{ph} is gas physical exergy flow rate,KW; E_{ch} is gas chemical exergy flow rate, KW.
Chemical exergy flow rate is calculated as:

IOP Conf. Series: Materials Science and Engineering **605** (2019) 012001 doi:10.1088/1757-899X/605/1/012001

$$\dot{E}_{ch} = \dot{n}(\sum_i x_i E_{0i} + RT_0 \sum_i x_i \ln x_i) \quad (3)$$

in this equation, $\dot{E}_{ch}$ is gas chemical exergy flow rate, KW; $\dot{n}$ is molar flow rate, kmol/s； E_{0i} is standard chemistry exergy of component i, KW; as shown in Table 2; R is the ideal gas constant 8.314KJ/(Kmol·K)； T_0 is the temperature in the reference state, K; x_i is the volume fraction or mole fraction of component i,%.

Table 2. Specific enthalpy,entropy and standard chemical exergy of components[10].

Gas	h_0 kJ · kmol^{-1}	s_0 kJ · kmol^{-1} · K^{-1}	E_{x0} kJ · kmol^{-1}
H_2	8468.00	130.57	236100
CO	8669.00	197.54	275100
CH_4		186.16	831650

The gas physical exergy flow rate is calculated as:

$$\dot{E}_{ph} = \dot{n}[(h - h_0) - T_0(s - s_0)] \quad (4)$$

Where h is the specific entropy,kJ/（mol·K） and s is the specific enthalpy at the given temperature and pressure,kJ/mol； h_0 is the standrd specific enthalpy and s_0 is the standrd specific entropy under the reference, its value is shown in Table 2;assuming that the gas is an ideal gas, the specific enthalpy difference and the specific entropy difference are calculated as follows:

$$\Delta h = h - h_0 = \int_{T_0}^{T} c_p dT \quad (5)$$

$$\Delta s = s - s_0 = \int_{T_0}^{T} \frac{c_p}{T} dT - R \ln \frac{p}{p_0} \quad (6)$$

Where p is the given pressure, pa; T is the given temperature, K; p_0 is the pressure under the reference state, pa; c_p is the specific constant pressure heat capacity;the specific constant pressure heat capacity is calculated as follows :

$$C_p = a' + b'T + c'T^2 + d'T^3 \quad (7)$$

The type of proportional coefficient according to Table 3 .

Table 3. Coefficients used in constant specific heat empirical equation[11].

Gas	a'	b'	c'	d'
H_2	29.11	-1.916×10^{-3}	4.003×10^{-6}	-8.704×10^{-10}
CO	28.16	1.675×10^{-3}	5.372×10^{-6}	-2.222×10^{-9}

CH_4	19.89	5.204×10^{-2}	1.269×10^{-5}	-1.101×10^{-8}

2.3.2. Gas energy efficiency. Gas is the main product of gasifier, so the effective use of gas energy is the gas exergy, exergy efficiency is calculated as

$$\eta_{fuel} = \frac{\dot{E}_{fuel}}{\dot{E}_{biomass} + \dot{E}_{steam}} \times 100\% \qquad (8)$$

in this equation, η_{fuel} is gas efficiency, %; $\dot{E}_{fuel}$ is gas exergy flow rate, KW; $\dot{E}_{biomass}$ is exergy flow rate of biomass, KW; $\dot{E}_{steam}$ is exergy flow rate of steam, KW.

*2.3.3. Calculation of low gas calorific value.*Come to its calculation formula[12]:

$Q_{LHV}=126CO+108H_2+359CH_4$

in the formula, Q_{LHV} is the gas calorific value,MJ/m³; CO, H_2 and CH_4 are the volume fraction of gas,%.

2.3.4. *Calculation formula of biomass exergy flow rate.*

$$\dot{E}_{biomass} = \dot{m}\,\beta_{biomass} L_{biomass} \qquad (9)$$

Where β is the correlation factor;$L_{biomass}$ is the low-grade heat value of biomass MJ / m³;m is the mass flow rate, kg / s; based on the literature [13-14], β is calculated as:

$$\beta_{biomass} = \frac{1.044 + 0.016H/C + 0.34930O/C(1+0.0531H/C) + 0.0493N/C}{1-041240/C} \qquad (10)$$

in the eqution, C, H, O and N are the volume fraction of elemental analysis in biomass,%.

3. Results and discussion

3.1. Influence of temperature

Figure 3 shows the TG curve at a heating rate of 10 K / min. From Fig. 3, it can be seen that the pyrolysis of the particles can be roughly divided into three stages. (1) At about 100°C, Differential thermogravimetry curve shows the first peak, at this point is mainly pine particles moisture evaporation stage. (2) It is obvious that the second peak at 340 °C is the pyrolysis stage of pine pellet, and the weight loss rate reaches the maximum at this time. It is the main phase of pyrolysis reaction and also the volatile release stage. (3) After the temperature is higher than 500 °C, the ash is mainly the further decomposition of the solid residue, finally reaching the weight loss rate of 19.06%. The result is consistent with the previous study of Niu [15].The results show that the production of gasification gas is mainly determined by one reaction, which mainly includes the reaction of volatilization and water vapor conversion[16-22].

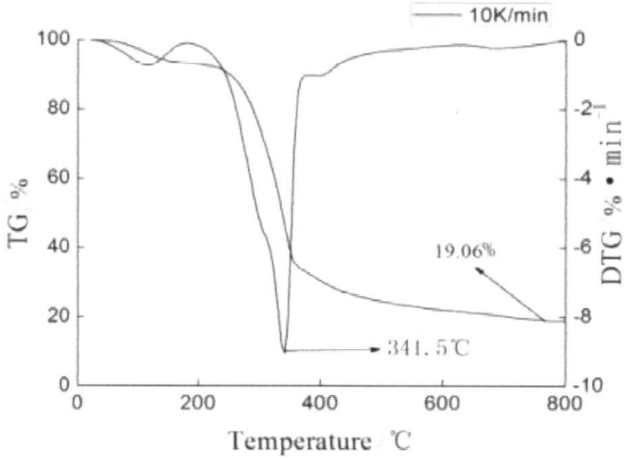

Figure 3. The pyrolysis weight loss curve of pine sawdust.

Figure 4 (a) and (b), respectively, at the steam flow of 0.3kg/h and 0.6kg/h, the gas volume fraction changes with temperature. It can be seen from Fig.4(a) that as the temperature increases, the volume fraction of H_2 increases from 27.45% to 38.38% with an increase of 10.93%,while the volume fraction of CO decreases and the minimum volume fraction reaches 22.21% at 800°C. On the one hand this is because more biomass and tar are cleaved by the elevated temperature and release more gas molecules, and on the other hand the water vapor reduction reaction ($C+2H_2O \rightarrow CO+2H_2$) is endothermic, increasing the temperature favors the forward reaction. With the increase of CO concentration, promoting the CO shift reaction ($CO+H_2O \rightarrow H_2+CO_2$) to the forward direction, which leads to the increase of H_2 volume fraction and the decrease of CO volume fraction. Franco et al. [23]. concluded that the steam reduction reaction ($C+2H_2O \rightarrow CO_2+2H_2$) became more dominant at a temperature higher than 830°C, similar to our experimental results. With a gradually increasing temperature, the rate of CO_2 reduction ($C+CO_2 \rightarrow 2CO$) begins to dominate, and the reaction is endothermic, resulting in an increase in the CO content of the gas. As the reaction of CH_4 with steam reforming is endothermic, with the increase of temperature, the reaction rate of methane ($CH_4+H_2O \rightarrow 3H_2+CO$) increases, and the volume fraction decreases from 17.29% to 11.98%. The trend of curves in Figure 4(a) are similar to Figure 4(b). When the temperature is 950 °C, the volume fractions of H_2 and CO increase from 38.38% and 27.67% of 0.3kg / h to 44.79% and 30.61% of 0.6kg / h, respectively, while CH_4 increases from 11.98 % To 9.94%, indicating that the appropriate increase in steam flow can get hydrogen-rich gas.

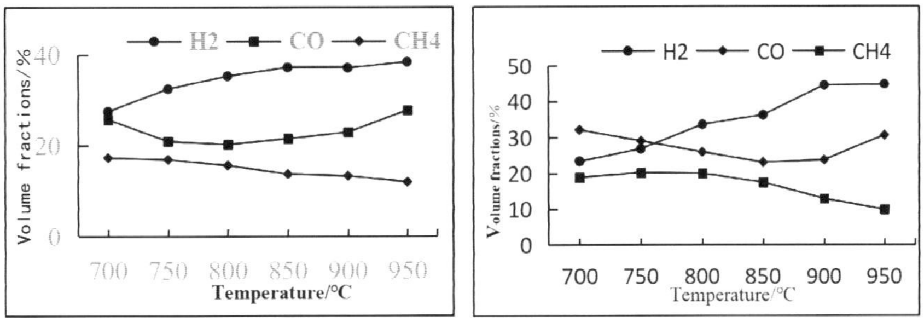

(a) (b)

Figure 4. Gas volume fraction changes with temperature with the different steam flow rate.

Figure 5(a) and Figure 5(b) respectively show the change of the total gas temperature with the steam flow rates of 0.3 kg/h and 0.6 kg/h. It can be seen from Figure 3 that with the increase of temperature, the gas total exergy increases from 10564.76KJ/kg at 700°C to 12483.38KJ/kg at 800°C and then decreases to 11026.82KJ/kg at 950°C, while the value gas group exergy are similar to the rules of Figure 4(a). The reason is that with the increase of temperature, the output of gas also increases, which leads to the increase of gas total exergy and gas group exergy. Figure 5(b) is similar to Figure 5(a) for the changing gas total exergy with temperature. At 800°C, the steam flow rate increases from 0.3kg/h to 0.6kg/h and the gas group exergy increases from 12483.38KJ/Kg to 40876.77KJ/Kg, with an increase of 28393.39KJ/Kg. Therefore, gasification effect at this temperature.

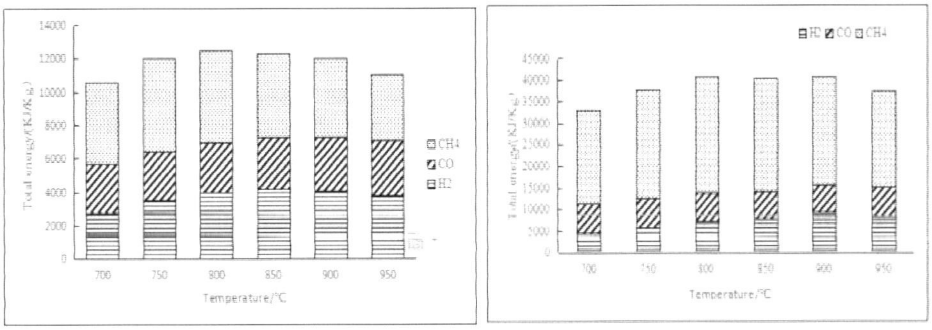

(a) (b)

Figure 5. The change of the total exergy of the gas at different temperature with the different steam flow rate.

Figure 6 shows the change of low gas calorific value with temperature at different steam flow rates. With different steam flow rates and temperatures, the low gas calorific value varies from 10 to 14MJ/m³, similar to that of Li [24].With a steam flow rate of 0.3kg/h, the low gas calorific value decreases with temperature, which is due to due to the relatively low calorific value of H_2. Although the volume fraction of H_2 is large, its increased calorific value can not make up for the decrease of calorific value of other hydrocarbon gases, so the caloric value of the gas is reduced. Figure 4(a) verifies this result. Steam flow increased from 0.3kg/h to 0.6kg/h, which enhanced the reforming reaction between water gas and CH_4, which resulted in the increase of gas production, which led to the increase of low gas calorific value at the corresponding temperature. [25-29]. At the temperature of 800°C, the low gas calorific value increased from 11965.36KJ/m^3 of steam flow rate of 0.3kg/h to 14076.72KJ/m^3 of 0.6kg/h, with an increase of 2111.36KJ/m^3, showing a positive impact on an increase of steam flow rate of gasification.

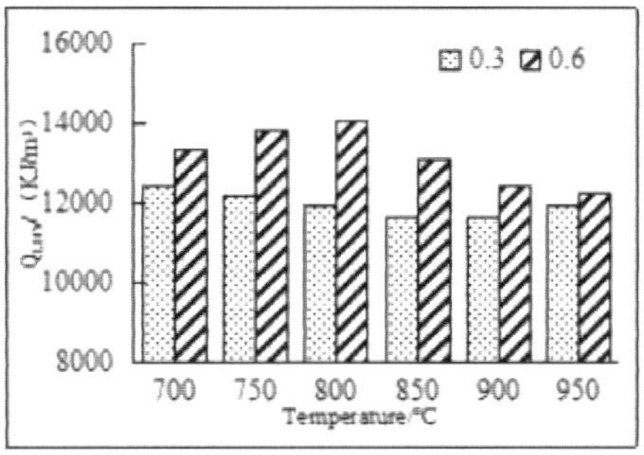

Figure 6. Low calorific value of gas changes with temperature in different steam flow rate.

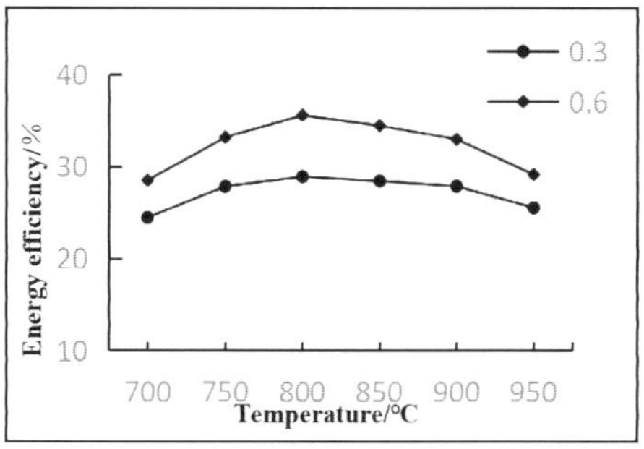

Figure 7. The change of gas exergy efficiency with temperature in different steam flow rate.

Figure 7 shows the change of gas exergy efficiency with temperature at different steam flow rates. It can be seen from the figure that as the temperature increases, the efficiency increases first and then decreases, and the change trend is similar to that in Figure 5(a) and Figure 5(b). With the increase of steam flow rate from 0.3kg/h to 0.6kg/h at 800°C, the gas exergy efficiency increased from 28.90% to 35.55%, an increase of 6.65% due to the increase of steam flow, the output of gas increases, so the efficiency of gas increases.

4. Conclusions

This article from the view of exergy about temperature and steam flow on the gas characteristics and gasification process performance indicators come to the following conclusions:

(1) At the steam flow rate of 0.3kg/h, the volume fraction of H_2 increases with the increase of temperature, the volume fraction of CO increases first and then decreases to 20.21 % at 800°C, the volume fraction of CH_4 decreases with temperature. When the steam flow rate increases to 0.6kg/h, the trend of the curves is similar to that of 0.3kg / h. Appropriately increasing the steam flow rate can produce hydrogen-rich gas. However, due to the change of gas volume fraction, the gas total exergy increased first and then decreased, increasing by 28393.39KJ / Kg at 800°C.

(2) At different steam flow rates, the low caloric value of gas varies from 10 to 14MJ/m^3; at 800°C, the steam flow rate increases from 0.3kg/h to 0.6kg/h, and the low calorific value have an increase of 2111.36KJ/m^3.

(3) The trend of the gas exergy efficiency with temperature is the same with the low gas calorific value at different steam flow rates as the temperature increases first and then decreases. At the highest exergy efficiency, the steam flow rate increased from 0.3kg/h to 0.6kg/h and the gas efficiency increased by 6.65%.

Acknowledgments

This work is supported by (1) National Natural Science Foundation of China, 51768054 (2) National Natural Science Foundation of China, 51764046 (3) Inner Mongolia Autonomous Region Education

Science "13th Five-Year Plan", NGJGH201708 (4) School-enterprise cooperation project, 2018073.

References

[1] Zhang X,Cai Z,Chen Y and Li H 2015 *J. Journal of Library and Information Sciences in Agriculture.* **37** 1-5
[2] Li W and Gao P 2013 *J.Environmental Science.* **15** 4-7
[3] Niu Y,Ma L J,Chen Y S and Chen J J 2015 *J.Transactions of the Chinese Society for Agricultural Machinery.* **46** 189-93
[4] Peterson D,Haase S 2009 *R. Market assessment of biomass gasification and combustion technology for small and medium-scale application.*NREL/TP-7A2-46190.
[5] Kong M,Fei J H,Wang S, Lu W and Zheng X 2011 *J. Bioresour Technol.***102** 2004
[6] Ptasinski K J,Hamelinck C and Kerkhof P J A M 2002 *J.Energy Conversion and Management.***43** 1445-57
[7] Zhan J H and You B 2009 *J.Nitrogenous Fertilizer Technology.***30** 3-4
[8] Dineer I.Technical 2002 *J.In-ternational Journal of Hydrogen Energy.***27** 265-285
[9] Prins MJ and Ptasinki KJ 2006 *J.Energy.***31** 3458-70
[10] Yuns A and Michael A 1994.*Thermodynamics : an Engineering Approach.*(New York : McGraw-Hill)
[11] Michael J Moran and Howard N Shapiro 2008.*Fundamentals of Engineering Thermodynamics.*(Hoboken: Wiley)
[12] Cohce M K,Dincer I and Rosen M A 2010 *J.International Journal of Hydrogen Energy.***35** 4970-80
[13] Ng Y C and Lipinski W 2012 *J.Energy.***44** 720-731
[14] Prins M J and Ptasinski KJ 2006 *J.Energy.***31** 3458-70
[15] Niu Y H, Han F T, Li Y K and Zhang X F 2017 *J. CIESC Journal.* **68** 1191-98.
[16] Ahmed I, Jangsawang W and Gupta A 2012 *J. Applied Energy.***91** 173-179
[17] Skoulou V, Kantarelis E, Arvelakis S,W. Yang and A. Zabaniotou 2009 *J.International Journal of Hydrogen Energy.* **34** 5666-73
[18] Kantarelis E, Donaj P, Yang W and A. Zabaniotou 2009 *J . Journal of Hazardous Materials.* **167** 675-684
[19] Donaj P, Yang W, Błasiak W and Christer F 2010 *J. Journal of Hazardous Materials.* **182** 80-89
[20] Donaj P, Blasiak W, Yang W and Christer F 2010 *J.Journal of Hazardous Materials.* **185** 472-481
[21] Zhou C, Stuermer T, Gunarathne R Weihong Y and Wlodzimierz B 2014 *J.Fuel.***122** 36-46
[22] Skoulou V, Swiderski A, Yang W and Zabaniotou A 2009 *J.Bioresource Technology.***100** 2444-51.
[23] Franco C,Pinto F and Gulyurtlu I 2003 *J. Fuel* .**82** 835-842
[24] Li L N,Ying H,Tu J L,Sun Y J and Xu Y 2011 *J.Chemistry and Industry of Forest Products.***31** 18-24
[25] Ma C R,Xiao B ,Chen Y M,Jiang J F and Zhou X M 2007 *J. Journal of Combustion Science and Technology.***13** 461-467
[26] SKOULOU V,SWIDERSKI A and YANG W 2009 *J. Bioresource Technology.***100** 2444-51

[27] Xie Y R,Xiao J and Shen L H 2008 *J.Acta Energiae Solaris Sinica.* **29** 888-893.
[28] GAO N B,LI A M and QUAN C 2008 *J. International of Hydrogen Energy.* **33** 5430-38.
[29] Wang Z W,Su X Y and Cheng C M 2008 *J.Journal of Fuel Chemstry and Technology.* **28** 298-305.

Performance Analysis of Single Cell Alkaline Electrolyser Using Mathematical Model

Y. Sanath K. De Silva, Peter Hugh Middleton, Mohan Kolhe

Dept. of Engineering Sciences, Faculty of Engineering and Science,

University of Agder (UiA), N-4898 Grimstad, Norway

sanath37@gmail.com, hugh.middleton@uia.no, mohan.l.kolhe@uia.no

Abstract. The generation of hydrogen using electrolysis process with integrated renewable energy sources is very important specially in environmental aspects. In this paper, we demonstrate that, the performance of electrolysis process which could be enhanced by decreasing the distance between electrodes while changing the properties of electrodes. At first, two single cell alkaline electrolysers are fabricated using an in-house 3D printer. Thereafter, the best performing cell is selected by considering its performance through different experiments. Finally, the performance of the selected cell is analysed by changing the distance between electrodes and changing the properties of electrodes. Throughout this research study the SolidWorks is used as the design software while using Mathlab Simulink as a modelling tool.

1. Introduction

Today, the generation of hydrogen using electrolysis process has been identified as a very important environmentally friendly method. The renewable energy source, particularly wind, hydro and solar can be integrated to generate hydrogen from electrolysis process by making the whole system as zero emission system. Thus, it is important to enhance the performance of the electrolysis cell in order to generate hydrogen through electrolysis process.

The performance of electrolysis cell is directly proportional to the properties of electrodes and the distance between electrodes. The ohmic loss of alkaline electrolysis cell is mainly induced by the electrical resistance of the electrolytes, which can be reduced by shortening the distance between the anode and cathode. If the distance between the anode and cathode is small, it may enhance the performance of the cell. Thus, there should be an optimum distance between the anode and cathode. Hence, the main aim of this research work is to investigate the optimum parameters to enhance the performance of electrolysis cell.

The manufacturing cost of an alkaline electrolysis cell is comparatively lower than other existing electrolysis process such as Proton Exchange Membrane (PEM) and Solid Oxide Electrolysis (SOE). The working principle of an alkaline electrolysis cell is shown in Fig. 1.

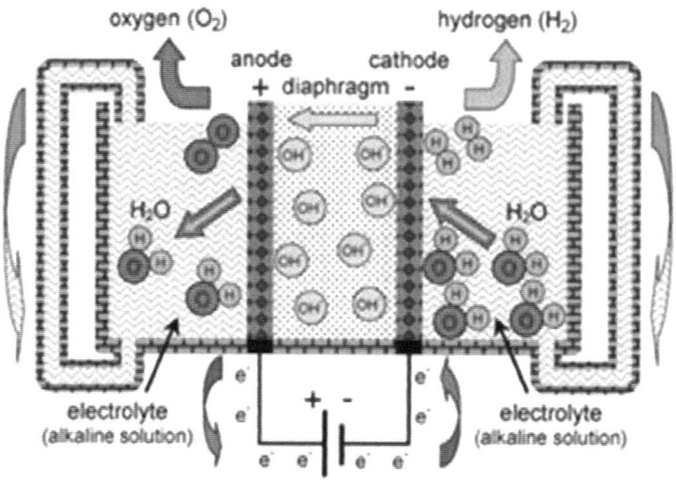

Figure 1. Working principle of single cell alkaline electrolyser [1].

The anode and cathode electrochemical reactions of alkaline electrolyser are illustrated as follows [2],

$$\text{Cathode:} \quad 2H_2O + 2e^- \rightarrow 2OH^- + H_2\uparrow \tag{1}$$

$$\text{Anode:} \quad 2OH^- \rightarrow H_2O + 2e^- + \frac{1}{2}O_2\uparrow \tag{2}$$

$$\text{Total:} \quad 2H_2O \rightarrow H_2 + \frac{1}{2}O_2\uparrow + H_2O \tag{3}$$

The generated hydrogen from the above-mentioned reactions do not require further rectifications since it has a purity level in between 95.5% - 99.9998% [3]. Furthermore, the electrical conductivity value of the feeding water should be below 5μS/cm in order to make sure the purity level of feeding water [4].

Some research studies are presented in the literature on alkaline electrolysers regarding design, modelling and different base materials which are used to enhance the performance of the cell. The different kind of electrolysis processes are integrated with renewable energy sources particularly wind, hydro and solar in Ref [5 – 9] to observe the performance of the system. In Ref [10-12], the static and dynamic model of alkaline electrolysis stacks are explained. Prior to that, the performance of electrolysis process can be enhanced by introducing new materials for electrolysis cell [13, 14]. In Ref [15], the ohmic resistance of alkaline electrolysis cell has been minimized through effective cell design. The zero-gap alkaline electrolysis cell design for renewable energy storage as hydrogen gas is discussed in Ref [16] in detail.

In this paper, we describe the design considerations of complete high-performance finite gap single cell alkaline electrolyser using 3D printed plastic housing since 3D printing technology is able to reduce the manufacturing cost of electrolysis cell by reducing costs for the post machining [17]. At the beginning of this study, two designs of electrolysis cells were fabricated and examined the respective performance, particularly the electrolysis power and efficiency of the cell were examined to select the best performing cell. Prior to that, the purity level and the flow rate of product gas were measured to compare the performance of fabricated cells. Thereafter, the performance of selected cell was enhanced through several experiments by changing the properties of electrodes and varying the distance between electrodes to reduce the ohmic loss of the cell. Throughout this research work, the Solidworks and MATHLAB/SIMULINK were used as design software and modelling tool respectively.

IOP Conf. Series: Materials Science and Engineering **605** (2019) 012002 doi:10.1088/1757-899X/605/1/012002

2. Methodology

In this research work, the main task can be classified into two groups as design and modelling of the cell, depending on the components and validation of the observed results as shown in Fig 2.

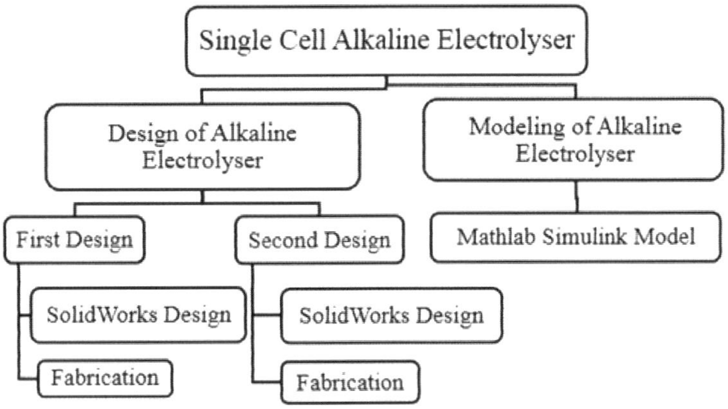

Figure 2. Classification of the design and model

SolidWorks software was used in the design part, to design base plates and the intermediate base plate. Thereafter, the lab-based 3D printer (Project type) was used to create a 3D print of designed cells. The printer is based on fused deposition modelling (FDM) 3D printing technology [18] and utilized a resin that is cured to form a dense plastic part. While fabricating the alkaline electrolyser, a sandwiching of all the inside parts occurs in between the base plates to implement the cell as shown in Fig. 3. At last, the observed results were validated by considering the results of mathematical model which was developed by MATLAB/SIMULINK modelling packages.

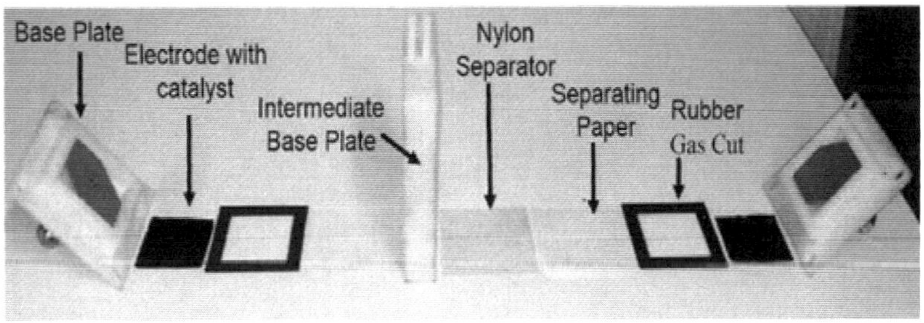

Figure 3. Basic parts of alkaline electrolysis cell

3. Design and fabrication of electrolysis cell

In this research study, two electrolysis cells were designed, and their performance was examined to identify the most suitable design. Figure 4 shows the first design of electrolysis cell and several experiments have been implemented by using first design to examine the performance of the cell. During the experiment with first design, it was observed that the product gasses contain some amount of water vapor by reducing the purity level of product gas. Thus, in order to enhance the purity level of product gas by eliminating this problem, the intermediate storage tank was designed on the intermediate base plate as shown in Figure 5 (second design). That storage tank was able to condense water vapor while increasing the purity level of product gas and the performance of the cell as well. Hence the second design was selected as the best performing electrolysis cell throughout this research study. Figure 6 shows the comparison of performance of the first design and the second design with the mathematical model.

When considering the first and second design, they are almost same in dimensions. Nevertheless, the only difference between first and second design is the intermediate storage tank which is mounted on the middle base plate in the second design.

Once the design process is over, the fabrication process of single cell alkaline electrolyser was started to obtain the required results. All the component parts of the single cell alkaline electrolyser are illustrated in Fig. 3.

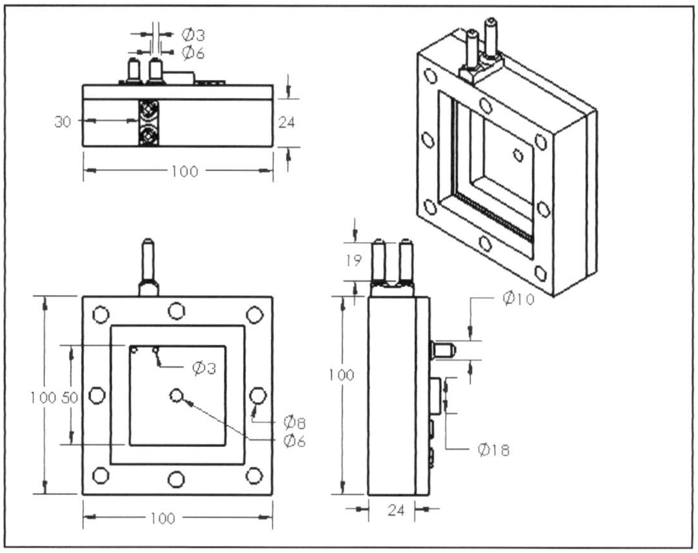

Figure 4. Detailed drawing of first design

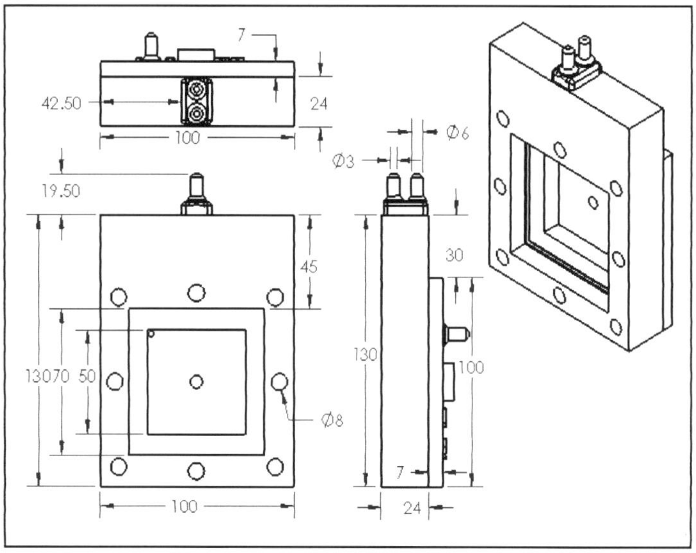

Figure 5. Detailed drawing of second design

According to the hydrogen evolution reaction (HER) and oxygen evolution reaction (OER) of the electrolysis process, nickel and its alloy can be used as electrodes of the cell [19], [20]. Thereafter the electrocatalyst was applied to electrodes in order to enhance the efficiency of the electrolysis process by reducing the activation energy for the electrochemical reaction. Furthermore, the catalyst layer consists of a Nafion® perfluorinated solution, 10% platinum on activated charcoal and methanol. The separating diaphragm (separating papers and nylon separator) separates the anode and cathode from each other.

IOP Conf. Series: Materials Science and Engineering **605** (2019) 012002 doi:10.1088/1757-899X/605/1/012002

The nylon mesh makes sure that there is no short circuit affects inside the cell. At last, to increase mechanical sealing inside the cell, the rubber was used as a gasket material.

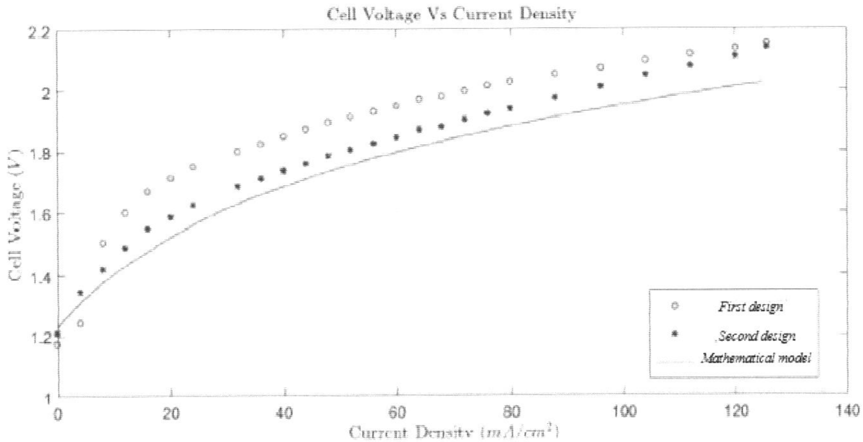

Figure 6. Performance comparison of first and second design with mathematical model

4. Modelling of electrolysis cell

The mathematical model of alkaline electrolysis cell is developed in MATHLAB/SIMULINK environment considering electrochemical equations. Figure 7 shows the V-I characteristic curve (Polarization Curve) of the typical alkaline electrolysis cell where the cell voltage is enhanced by some primary sources that caused to reduce the efficiency of the cell [21]. Hence, the cell voltage is a function of the overvoltages and reversible voltage as defined in (4).

$$V_{cell} = V_{rev} + V_{act} + V_{ohm} + V_{con} \tag{4}$$

Where, V_{rev} = reversible cell voltage, V_{act} = activation overvoltage, V_{ohm} = ohmic losses, V_{con} = concentration overvoltage.

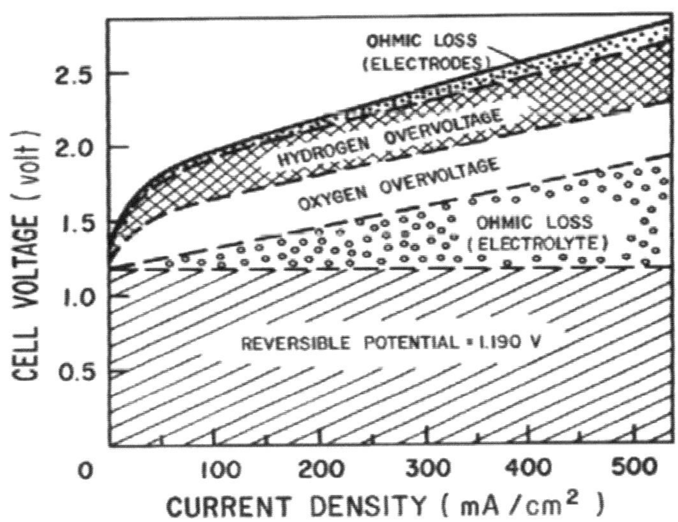

Figure 7. V-I Characteristic curve of typical electrolysis cell [22]

The V_{rev} can be calculated using the Gibbs equation as illustrated in (5) [23].

$$\Delta G = zFV_{rev} \tag{5}$$

Where, ΔG is Gibbs energy (237.2 kJ mol−1) [20] and F is Faraday's Constant (96485 C mol−1) [23]. The analytical value of z equals to 2. Moreover, the V_{act} can be defined by following expression in (6).

$$V_{act} = s\log\left\{\frac{t_1 + \frac{t_2}{T} + \frac{t_3}{T}}{A} I + 1\right\} \tag{6}$$

Where s, t_1, t_2, t_3 are the coefficients for overvoltage on electrodes, I is the current density, T is the temperature of the cell and A is the electrode area respectively [23], [24].

The electrical resistance of the electrolytes causes to generate the ohmic loss (V_{ohm}) which mitigate the performance of the cell by a considerable amount. However, for the optimum anode and cathode distance, the electrolysis cell is able to reduce the ohmic loss by increasing the cell performance. This overvoltage can be determined as in (7) [21].

$$V_{ohm} = \frac{r_1 + r_2 T}{A} I \tag{7}$$

Where r_1, r_2 = ohmic resistance parameters [23], [24]. The V_{con} is neglected in this modelling part since it has extremely low value compared to the V_{act} and V_{ohm}. Hence, the equation (4) can be rewritten as in (8).

$$V_{cell} = V_{rev} + V_{act} + V_{ohm} \tag{8}$$

4.1. Simulation background

Table 1 illustrates the relevant parameters including units which is used in this mathematical modelling study. Reference [23] and [24] are explained in the modelling part of electrolysis cell and most of the data available in the table are taken from that research article.

Table 1. Details of constants for mathematical modelling

Name of Constants	Symbols	Units	Value
Reversible Voltage	V_{rev}	V	1.229
Area of Electrode	A	cm^2	0.25
Faraday's Constant	F	$Cmol^{-1}$	96485
Number of Electrons	Z		2
Coefficient for overvoltage on electrodes	s	V	0.185
	t_1	$A^{-1}m^2$	1.002
Coefficient for overvoltage on electrodes	t_2	$A^{-1}m^2\,^0C$	8.424
	t_3	$A^{-1}m^2\,^0C$	247.3
Parameter related to ohmic resistance of electrolyte	r_1	Ωm^2	$8.05e^{-5}$
	r_2	$\Omega m^2\,^0C^{-1}$	$-2.5e^{--7}$

4.2. Mathematical model of electrolysis cell

A developed mathematical model in MATLAB/SIMULINK environment, is illustrated in Figure 8. The operating temperature (T) and area of the cell (A) are selected as inputs during this modelling study.

The V-I characteristic and P-I characteristic of alkaline electrolyser obtained by mathematical model are shown in Figure 9 and Figure 10 respectively. Furthermore, the cell voltage and overvoltage characteristics are illustrated in Figure 11.

IOP Conf. Series: Materials Science and Engineering **605** (2019) 012002 doi:10.1088/1757-899X/605/1/012002

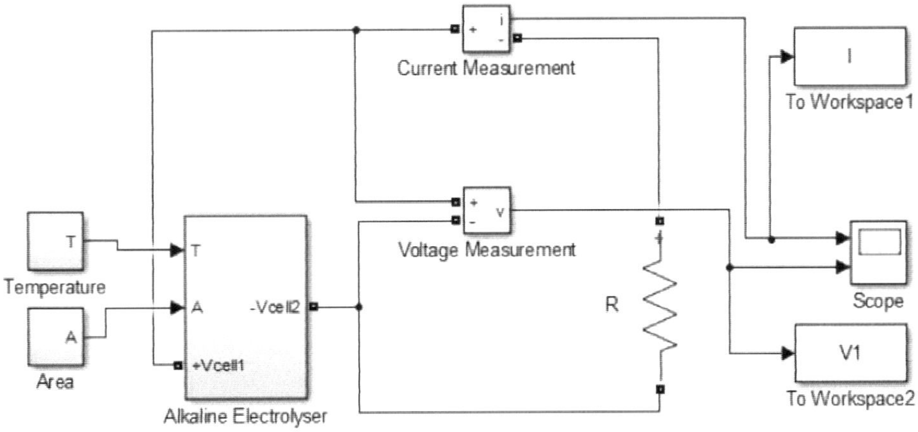

Figure 8. Mathematical model of alkaline electrolysis cell in MATLAB/SIMULINK

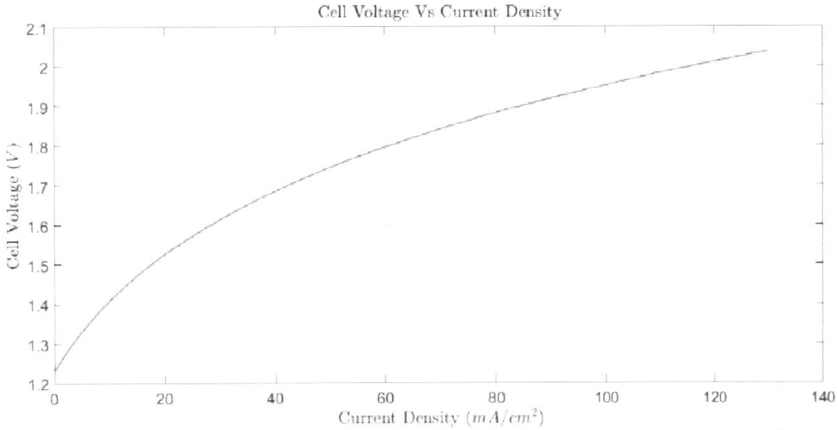

Figure 9. V-I characteristic of alkaline electrolysis cell

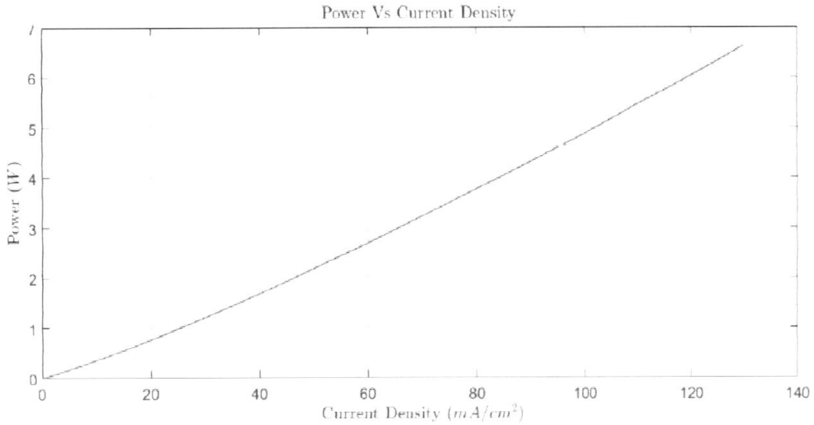

Figure 10. P-I characteristic of alkaline electrolysis cell

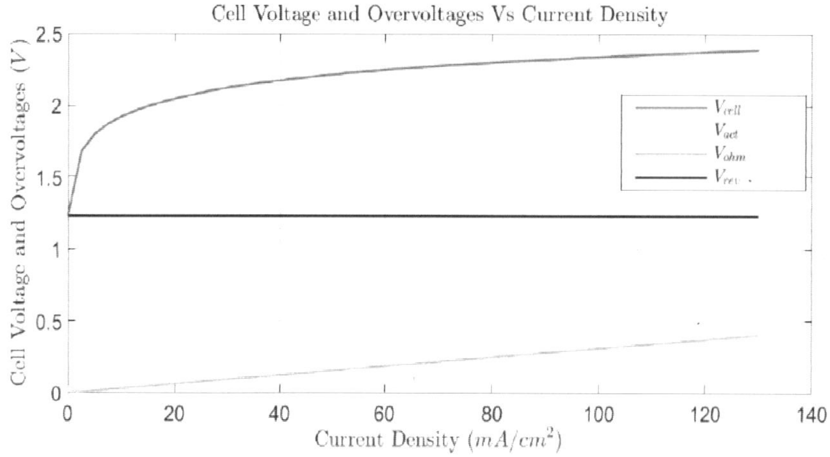

Figure 11. Cell voltage and overvoltages of alkaline electrolysis cell

5. Experiments and results

Different experiments have been done in order to enhance the performance of single cell alkaline electrolyser by changing the properties of Membrane Electrode Assembly (MEA) and electrode distance of the cell.

5.1. Experiment I: Effect of Distance Between Electrodes

The distance between electrodes of electrolysis cell is a critical fact to make deviation of the electrolysis performance, particularly power and efficiency. The performance of electrolysis cell could be increased by keeping optimum distance between the anode and cathode while reducing the ohmic loss inside the cell.

In order to investigate the effect of electrode distance for electrolysis process, three experiments were implemented while varying the distance between electrodes as 2.5mm, 1.9mm and 1.3 mm respectively. Figure 12 shows the bench setup of the Experiment I. The V-I characteristic curves of each experiment were drawn with the mathematical model by considering experimental data to examine the performance of the cell as shown in Figure 13. According to the observation of this experiment, 1.3 mm thick electrode assembly was exhibited better performance compared with other two experiments.

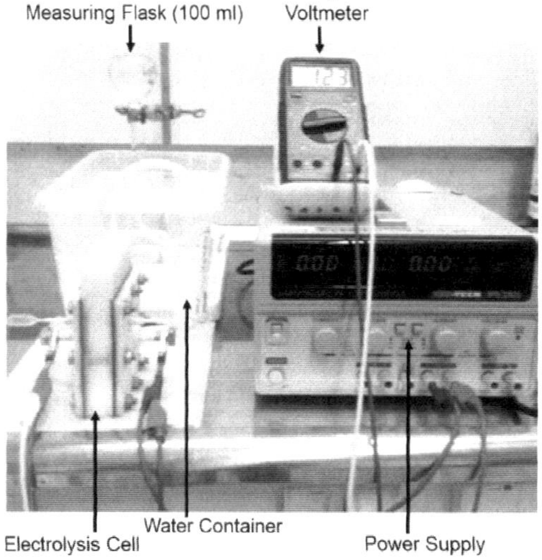

Figure 12. Bench setup of the Experiment I

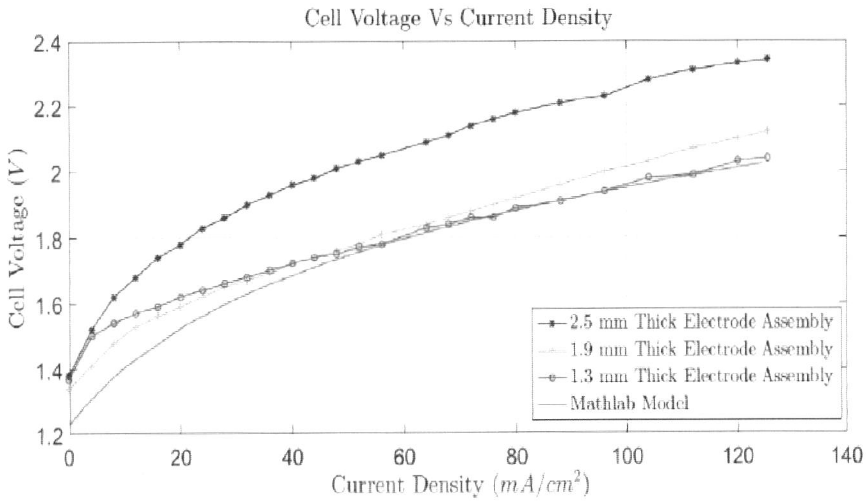

Figure 13. Comparison of the Experiments (Electrode gap 2.5mm, 1.9 mm, 1.3mm) with the mathematical model

5.2. Experiment II: Electrolysis Cell with Fine Nylon Mesh

In this experiment, one nylon mesh and two filter papers were adopted as separating diaphragm and, the same procedure was continued to observe required data by keeping the gap between electrodes as 1.3 mm. Table 2 and 3 shows the observed data and calculated parameters for Experiment II.

Table 2. Flow rate calculation data of the Experiment II

Produced Gases	Cell Current (A)	Voltage of Cell (V)	Time (s)
Hydrogen ($H_2(g)$)	2.5	2	334
	3.1	2.05	254
Oxygen ($O_2(g)$)	2.5	2	613
	3.14	2.05	551

Table 3. Calculated parameters of the Experiment II

Flow rate calculation		Power and efficiency calculation	
Variable	Value	Variable	Value
Current of the cell (A)	3.1	Voltage of cell (V)	2.05
n_{H_2} (mol/min)	9.6388 x 10^{-4}	P_{Total} (W)	6.355
n_{O_2} (mol/min)	4.8194 x 10^{-4}	R_{Ohmic} (Ω)	0.2
$V_{H_2}(g)$ (Theo)(cm^3/min)	23.71	P_{Ohmic} (W)	1.922
$V_{O_2}(g)$ (Theo)(cm^3/min)	11.86	$P_{electrolysis}$ (W)	4.433
$V_{H_2}(g)$ (Exp)(cm^3/min)	23.62	$\eta_{electrolysis}$ (%)	69.8
$V_{O_2}(g)$ (Exp)(cm^3/min)	10.89	η_{energy} (%)	72.2

5.3. Experiment III: Effect from the Thickness of Electrodes

In this experiment, the thickness of the electrodes was enhanced by adding porous nickel plates to MEA. Same bench setup was deployed as in Experiment 1 and observe the required data for efficiency and power calculations. The performance comparison was done through the V-I characteristic curve which is generated from a mathematical model.

Table 4. Flow rate calculation data of the Experiment III

Produced Gases	Cell Current (A)	Voltage of Cell (V)	Time (s)
Hydrogen ($H_2(g)$)	2.5	2.03	301
	3.14	2.1	260
Oxygen ($O_2(g)$)	2.5	2.06	672
	3.14	2.1	554

Table 5. Calculated parameters of the Experiment III

Flow rate calculation		Power and efficiency calculation	
Variable	Value	Variable	Value
Current of the cell (A)	3.14	Voltage of cell (V)	2.1
n_{H_2} (mol/min)	9.7632 x 10^{-4}	P_{Total} (W)	6.594
n_{O_2} (mol/min)	4.8816 x 10^{-4}	R_{Ohmic} (Ω)	0.207
$V_{H_2}(g)$ (Theo)(cm^3/min)	24.017	P_{Ohmic} (W)	2.0433
$V_{O_2}(g)$ (Theo)(cm^3/min)	12.008	$P_{electrolysis}$ (W)	4.5507
$V_{H_2}(g)$ (Exp)(cm^3/min)	23.077	$\eta_{electrolysis}$ (%)	69.01
$V_{O_2}(g)$ (Exp)(cm^3/min)	10.83	η_{energy} (%)	70.5

5.4. Experiment IV: Electrodes with Porous Stainless-Steel Plates

To reduce the ohmic overvoltage, electrodes of the cell were replaced by porous stainless-steel plates and the relevant data was observed as in previous experiments by keeping the electrode gap as 1.3 mm. The observed data and calculated parameters are illustrated in Table 6 and 7 respectively.

Table 6. Flow rate calculation data of the Experiment IV

Produced Gases	Cell Current (A)	Voltage of Cell (V)	Time (s)
Hydrogen ($H_2(g)$)	2.5	1.98	300
	3.1	2.04	250
Oxygen ($O_2(g)$)	2.5	1.98	588
	3.14	2.04	505

Table 7. Calculated parameters of the Experiment IV

Flow rate calculation		Power and efficiency calculation	
Variable	Value	Variable	Value
Current of the cell (A)	3.14	Voltage of cell (V)	2.05
n_{H_2} (mol/min)	9.7632 x 10^{-4}	P_{Total} (W)	6.4056
n_{O_2} (mol/min)	4.8816 x 10^{-4}	R_{Ohmic} (Ω)	0.194
$V_{H_2}(g)$ (Theo)(cm^3/min)	23.017	P_{Ohmic} (W)	1.914
$V_{O_2}(g)$ (Theo)(cm^3/min)	12.01	$P_{electrolysis}$ (W)	4.4921
$V_{H_2}(g)$ (Exp)(cm^3/min)	24	$\eta_{electrolysis}$ (%)	70.13
$V_{O_2}(g)$ (Exp)(cm^3/min)	11.88	η_{energy} (%)	72.5

Figure 14 is illustrated combinations of V-I characteristic curves of the Experiments II, III and IV with the mathematical model as a comparison of particular experiments.

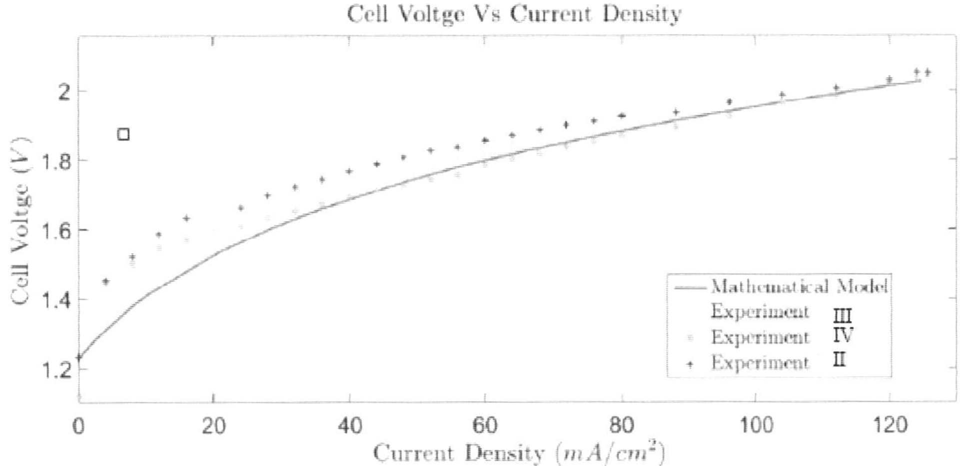

Figure 14. Comparison of the Experiments II, III, IV with the mathematical model

5.5. Experiment V: Electrolysis Cell with Polymeric Membrane Separator

In this experiment Nafion 115 and Nafion 117 polymeric membranes were deployed as a separating diaphragm respectively. According to the observed data, the Nafion 115 was given a good performance compared to the Nafion 117. Nevertheless, the performance of both polymeric membranes is much lower than when compared with other experiments. The performance comparison of both polymeric membranes with a mathematical model is shown in Figure 15.

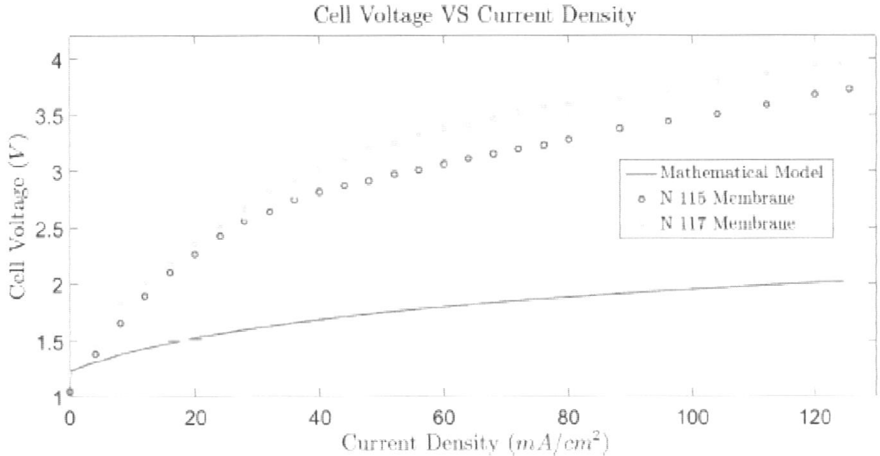

Figure 15. Performance comparison of both polymeric membranes with the mathematical model

6. Conclusion

In this research study, we have enhanced the performance of the electrolysis process by decreasing the distance between the anode and cathode while changing the properties of electrodes. The main aim of this research was to investigate the optimum parameters to enhance the performance of electrolysis cell using mathematical model. Thus, two designs of single cell alkaline electrolysis cell were deployed, and second design was selected as the best performing cell after considering their performance separately. The Experiment IV is selected as the best performing cell among five Experiments after considering the performance comparison graph with the mathematical model. In that experiment the distance between porous electrodes was set to 1.3 mm to reduce the ohmic overvoltage inside the cell by increasing the efficiency and power. According to the observation of Experiment IV, the practical values of hydrogen gas flow rate, oxygen gas flow rate and electrolysis efficiency are calculated as 24 cm3/min, 11.88 cm3/min and 70.13% respectively. The nafion is an acid and react with alkaline solution by reducing the lifetime of the cell. Nevertheless, the life expectancy of the cell is obviously adequate for experimental purposes. In the future, we intend to develop a zero-gap electrolysis cell with different catalyst layers to reduce the ohmic loss furthermore and to integrate optical sensing technology to observe the optimum distance between electrodes quite accurately. During this research study the SolidWorks and MATLAB/Simulink were used as design and modelling tools respectively. As a whole, it is obvious that by shortening the distance between the anode and cathode, the performance of electrolysis cell can be enhanced by a considerable amount in terms of electrolysis efficiency and power.

References

[1] A. Ursua, L. M. Gandia, and P. Sanchis, "Hydrogen production from water electrolysis: Current status and future trends," Proceedings of the IEEE, vol. 100, no. 2, pp. 410-426, Feb. 2012.
[2] K. Zeng and D. Zhang, "Recent progress in alkaline water electrolysis for hydrogen production and applications," Progress in Energy and Combustion Science, vol. 36, no. 3, pp. 307-326, Jun. 2010.
[3] Johanna Ivy, "Summary of electrolytic hydrogen production," Milestone Completion Report, 2004.
[4] L. Bertuccioli, A. Chan, D. Hart, F. Lehner, B. Madden, and E. Standen, "Study on development of water electrolysis in the EU," Fuel Cells and Hydrogen Joint Undertaking, p. Final Report, Feb. 2014.
[5] A. Ursa, I. San Martn, E. L. Barrios, and P. Sanchis, "Stand-alone operation of an alkaline water electrolyser fed by wind and photovoltaic systems," International Journal of Hydrogen Energy, vol. 38, no. 35, pp. 14952-14967, Nov. 2013.
[6] M. Kiaee, P. Chladek, D. Infield, and A. Cruden, "Utilisation of alkaline electrolysers to improve power system frequency stability with a high penetration of wind power," IET Renewable Power Generation, vol. 8, no. 5, pp. 529-536, Jul. 2014.
[7] W. Jiang, et al." Identification and Power Electronic Module Design of a Solar Powered Hydrogen Electrolyzer," IEEE Asia-Pacific Power and Energy Engineering Conference, pp. 1-4. 2012.
[8] M. Kiaee, A. Cruden, D. Infield, Y. Ma, T. G.Douglas, "The impact on the electrical grid of hydrogen production from alkaline electrolysers," IEEE Universities Power Engineering Conference (UPEC), 45th International, pp. 1-6. 2010.
[9] N. Gyawali and Y. Ohsawa, "Integrating fuel cell/Electrolyzer/Ultracapacitor system into a stand-alone Microhydro plant," IEEE Transactions on Energy Conversion, vol. 25, no. 4, pp. 1092-1101, Dec. 2010.
[10] M. ul-Karim1, M. T. Iqbal,"Dynamic modeling and simulation of alkaline type electrolyzers," Electrical and Computer Engineering, CCECE'09. IEEE Canadian Conference, pp. 711-715. 2009.
[11] K. Belmokhtar, M. L. Doumbia, K. Agbossou,"Dynamic model of an alkaline electrolyzer based an artificial neural networks," Ecological Vehicles and Renewable Energies (EVER), IEEE 8th International Conference and Exhibition, pp. 1-4. 2013.
[12] I. D. de Cerio Mendaza and B. Bak-Jensen,"Alkaline electrolyzer and V2G system DIgSILENT models for demand response analysis in future distribution networks," PowerTech (POWERTECH),IEEE Grenoble, pp. 1-8. 2013.
[13] A. Cruden, D. Infield, M. Kiaee, T. G. Douglas, and A. Roy, "Development of new materials for alkaline electrolysers and investigation of the potential electrolysis impact on the electrical grid," Renewable Energy, vol. 49, pp. 53-57, Jan. 2013.
[14] Ph. Vermeiren, W. Adriansens, J. P. Moreels, R. Leysen, "Evaluation of the ZirfonR separator for use in alkaline water electrolysis and Ni-H 2 batteries," International Journal of Hydrogen Energy, 23.5: 321-324.1998.
[15] R. Phillips, A. Edwards, B. Rome, D. Jones, and C. Dunnill, "Minimising the ohmic resistance of an alkaline electrolysis cell through effective cell design," International Journal of Hydrogen

Energy, 42(38), pp.23986-23994, 2017.

[16] R. Phillips and C. Dunnill, "Zero gap alkaline electrolysis cell design for renewable energy storage as hydrogen gas," RSC Advances, 6(102), pp.100643-100651, 2016.

[17] Y. Sanath K. De Silva, M.J.M.A. Rasul, P. Middleton and M. Kolhe," PEM FC Single Cell Based on a 3-D Printed Plastic Housing and Experimental Validation with the Mathematical Model", Energy Procedia, 144, pp.63-74, 2018.

[18] Usglobalimages.stratasys.com. (2018). [online] Available at: http://usglobalimages.stratasys.com/Main/Files/Machine_Spec_Sheets/PSS_FDM_Dim1200es.pdf [Accessed 15 May 2018].

[19] Md Mamoon Rashid, Mohammed K. Al Mesfer, Hamid Naseem, Mohd Danish. Hydrogen production by water electrolysis: a review of alkaline water electrolysis, pem water electrolysis and high temperature water electrolysis," International Journal of Engineering and Advanced Technology (IJEAT), vol. Volume-4, no. Issue-3, pp. 2249-8958, Feb. 2015.

[20] M. Wang, Z. Wang, X. Gong, and Z. Guo, "The intensification technologies to water electrolysis for hydrogen production - A review," Renewable and Sustainable Energy Reviews, vol. 29, pp. 573-588, Jan. 2014.

[21] A. Ursa and P. Sanchis," Static-dynamic modelling of the electrical behavior of a commercial advanced alkaline water electrolyser," International Journal of Hydrogen Energy, vol. 37, no. 24, pp. 18598-18614, Dec. 2012

[22] Isao. Abe, "Alkaline Water Electrolysis". Energy Carriers And Conversation Systems - Vol. 1. Chiba, Japan: Encyclopedia of Life Support System (EOLSS)

[23] A. S. Tijani, N. A. B. Yusup, and A. H. A. Rahim, "Mathematical Modelling and simulation analysis of advanced alkaline Electrolyzer system for hydrogen production," Procedia Technology, vol. 15, pp. 798-806, 2014.

[24] O. Ulleberg, "Modeling of advanced alkaline electrolyzers: A system simulation approach," International Journal of Hydrogen Energy, vol. 28, no. 1, pp. 21-33, Jan. 2003.

IOP Conf. Series: Materials Science and Engineering **605** (2019) 012003 doi:10.1088/1757-899X/605/1/012003

Hydrodynamic and gasification studies of sawdust in a pressurized circulating fluidized bed

Abinash Mahapatro*, Pinakeswar Mahanta

Department of Mechanical Engineering, Indian Institute of Technology Guwahati Guwahati, Assam India-781039

abinashlipun@iitg.ac.in

Abstract. The present research is focused on the hydrodynamics and gasification studies of a pressurized circulating fluidized bed (PCFB) gasifier considering saw dust as a bed inventory. The operating pressure range of the PCFB gasifier is maintained in between 1 to 4 bar and the suspension density, syngas composition, lower heating value (LHV), gas yield, carbon conversion efficiency (CCE) and cold gas efficiency (CGE) were calculated and presented in this paper. The suspension density increases with rise in operating pressure and decreases with rise in superficial velocity. The concentration of CH_4 in the syngas was found to be increasing with rise in bed pressure and is found to be 4.78 % at 4 bar. However, no significant effect of operating pressure on H_2 concentration is observed and is found to vary between 8.3 to 7.95 %. The CO concentration increases up to 2 bar pressure and decreases gradually afterwards. The LHV, gas yield, CCE and CGE has a positive response to the rise in operating pressure and found maximum at 4 bar. The CCE and CGE of the PCFB gasifier are increased by 57 % and 58 %, respectively when the operating pressure is changed from 1 to 4 bar.

Keywords: Gasification; Sawdust; Pressurized circulating fluidized bed

1. Introduction

The worldwide energy consumption is increasing rapidly due to the development of advanced power generation technologies in the last two decades because of which the global community is suffering with fuel crises, global warming and pollutant emission from the sources of conventional fossil fuel. These serious effects attract researchers for an intensive research in field of clean energy for the next generation. Solar, biomass, wind, tidal and other renewable energies act as a bridge between environmental pollution and clean energy. Biomass is preferred over all other types of renewable resources of energy due to its control in hand, easy to transport like coal and storage in the stockyard. It has also gained popularity after coal, petrochemicals and natural gas with 14 % of the global share in energy consumption [1]. In a tropical country like India, a huge amount of biomass are stored in an un-hygienic condition which are under-utilized. The low calorific value and minimal ash content of biomasses attract investigators for the production of power in an efficient way [2-4]. Fluidized bed is one of the promising technology because of its capacity to handle a wide range of fuel and uniform-mixing of particles to enhance gasification rate inside the bed [5].

The bed hydrodynamics of the circulating fluidized bed (CFB) riser plays a vital role to investigate the different fluidization regime and design and operation of combustion/gasification. It shows a balance of pressure and mass of the circulating loop and classifies the solid distribution in the axial direction of the riser section [6-7]. Hydrodynamic study includes the effects of superficial velocity, particle size, solid inventory and operating pressure in the bed voiadge and suspension density of the riser [8]. Kalita et al. [9] studied the bed hydrodynamics, heat transfer behaviour and effects of solid inventory, particle size, and operating pressure considering blending of sawdust as a bed material.

The bubbling fluidized bed (BFB) gasification is the most preferred technology used by many researchers because of its ease of handling. Datta et al. [10] described the agglomeration behaviours of the high ash Indian coal gasification. Kumari and Vairakannu [11] studied the underground coal gasification of high and low ash Indian coal. Kumar et al. [12] also experimentally studied the air and steam mixtures in a bubbling gasifier and reported that the efficiency increases with decrease in air-fuel ratio. Miccio et al. [13] used the German brown coal and woodchips as biomass in dual fluidized bed for co-gasification at atmospheric conditions. Co-gasification of coal and biomass in a bubbling fluidized bed gasifier was also investigated by Li et al. [14]. They reported that H_2 and CO decreases with increase in equivalence ratio and increases with biomass ratio. Mallick et al. [15] studied the effect of equivalence ratio on concentration of H_2 and CO and found similar results as observed by Li et al. [14].

The atmospheric fluidized bed gasification is an established technology while pressurized fluidized bed gasification is still in very young age in research [16-17]. Gul et al. [18] studied the pressurised bubbling gasification and the syngas composition under different mass flow rates with an additional CO_2 reagents. Huang et al. [19] investigated the gasification of coal in a pressurized bubbling fluidized bed and reported the increase of gasification with increase in pressure. They also reported syngas composition to be a function of temperature but had a little effect of pressure on syngas composition with change of pressure from range 0.5 to 1.4 MPa. Srinivas et al. [20] studied the thermodynamic equilibrium model for a PCFB gasifier. They reported that the syngas concentration of H_2 and CO_2 decreased and CH_4 and CO increased with increases in gasifier pressure. They also reported that the effect of pressure on gas composition was very little while it affected the temperature, LHV of gas and exergy efficiency. Kitzler et al. [21] explored the pressurized gasification in a bubbling fluidized bed considering woddy biomass as a bed inventory. They reported that concentration of H_2 was higher if the gasification temperature was greater than 820^0 C and the pressure was less than 4 bar. The concentration of H_2 in a pressurized bubbling fluidized bed gasification taking sawdust as an inventory was reported by Long et al. [22]. They found an increase of concentration of H_2 and CH_4 and a decrease of CO_2 and CO with the increases in gasifier pressure from 1 to 4 bar. Berrueco et al. [23] investigated the effect of temperature and dolomite on formation of tar and gas concentration in a pressurized fluidized bed at constant pressure of 0.5 MPa. They reported that with introduction of dolomite as a bed material instead of sand, there was an increase of syngas concentration of H_2, CO, CH_4 and CO_2 with the increase in temperature. Sanchez et al. [24]. studied a new model for coal gasification in a bubbling pressurized fluidized bed, and reported the pressure to have a significant effect on gas composition. Also, there was a decrease in H_2 and CO while concentration of CH_4 and CO_2 increased with pressure.

It is observed that a very limited study has been reported regarding utilization of sawdust in PCFB. Further, PCFB has gained importance in recent time due to its compactness and high efficiency. Therefore, the current study is an attempt to investigate the hydrodynamic behaviour and gasification study of PCFB with sawdust as a bed inventory. The variation of suspension density, bed temperature, gas concentration, LHV, gas yield, CCE and CGE are studied under various operating conditions viz. pressure and superficial velocity.

2. Experimental setup and procedure

The schematic diagram of the PCFB unit is presented in Fig. 1. The PCFB unit is comprised of various components such as a riser, a distributor plate, a cyclone separator, a downcomer. The riser is fabricated with ID (internal diameter) of 100 mm, OD (outer diameter) of 110 mm and height of 3000 mm, which is further connected to cyclone separator through a connecting pipe. The cyclone separator is fabricated from stainless steel having dimensions of 660 mm in length and having barrel ID of 160 mm. The top part of the cyclone separator is used for collection of syngas where the bottom end is connected to the downcomer of ID 50 mm. Subsequently, the return leg is connected to the bottom of the riser column at 150 mm, above the distributor plate. A gateway valve is installed in the downcomer section to avoid any backflow from the riser. The pressure drop along the height of the riser is measured by using ten pressure sensors above the distributor plate and suspension density is calculated at those points. Similarly thirteen thermocouples are mounted along the height of the riser from distributor plate to monitor the bed temperature.

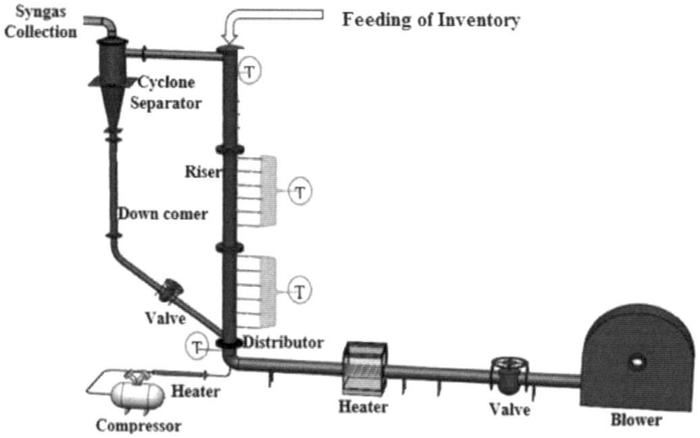

Figure 1. Experiment set up of hot bed PCFB unit.

For the atmospheric circulating fluidized bed (ACFB) experiments, the blower delivery air passes through an electric heater box before entering into the bottom of the riser through the distributor plate. On the other hand, for the PCFB experiments, a reciprocating compressor is used for pressurizing the air, which passes through a tubular heater before entering to the riser through distributor plate. The atmospheric/pressurized air is further routed through a straight-hole based distributor plate having opening area of 16.4%. A mass flow meter is installed for calculating the mass flow rate of the air that is entering into the riser.

Initially, the sawdust is sieved into particle size of 600 μm and further desired quantity of inventory is fed into the PCFB unit through the top of the riser. The air from the compressor passes through a heater before enrooting into the distributor plate. To regulate the airflow rate from the blower and compressor, a valve is installed. The temperature in the riser increases due to the burning of the sawdust and takes almost 10-20 minutes to achieve steady state flow regime in the riser during hot bed studies. In the meantime, the syngas is collected through the top of the cyclone separator. Furthermore, the gas chromatography is conducted to find out the quality of the collected syngas. After 30 minutes, the sawdust burns out in the riser and the temperature and the quality of syngas decrease. During the hot bed studies, experiments are conducted both at atmospheric as well as pressurized conditions. All the readings are recorded throughout the experiments with the help of the data acquisition system.

The experimental uncertainty is determined for suspension density by considering all the parameters. The parameters affect the suspension density are bed voidage and density of both air and sawdust. While calculating the bed voidage, the error during the measurement of pressure drop via pressure sensor is also taken into consideration. For suspension density, the overall uncertainty in the experimentation is found to be $\pm 3.25\%$ [25].

3. Results and discussion

The sawdust collected locally is used as an inventory for the present experiments. The properties of sawdust is same as those given by Mallick et al. [15]. The suspension density along the height of the riser, and effect of pressure on gasification studies i.e. effect of temperature, gas composition, LHV, gas yield, CCE and CGE is described in subsequent sections for the batch type study in a PCFB gasifier.

3.1. Effect of pressure and velocity on suspension density with saw dust as inventory

The variation of suspension density with pressure along the riser height at different superficial velocity for 1000 gm saw dust is shown in Figs. 2 through 4. The suspension density seems to be increasing as the pressure increases at a particular superficial velocity and also decreases with an increases in superficial velocity. As perceived from the figures, suspension density is having higher magnitude at the lower section of the riser height, which then, decreases up to the top middle zone, and then starts increasing to the exit of the riser. The maximum suspension density is observed in case of 4 bar operating at superficial velocity of 3 m/s. In view of deducing the effect of superficial velocity at fixed operating pressure, the maximum suspension densities magnitudes have been compared. This shows a decrement of 25 % and 41 % for 4 m/s and 5 m/s superficial velocity, respectively in comparison to 3 m/s condition. For the hydrodynamics study, the optimal operating conditions are superficial velocity = 3 m/s, and operating pressure = 4 bar where the magnitude of suspension density is higher i.e., 12.33 kg/m^3.

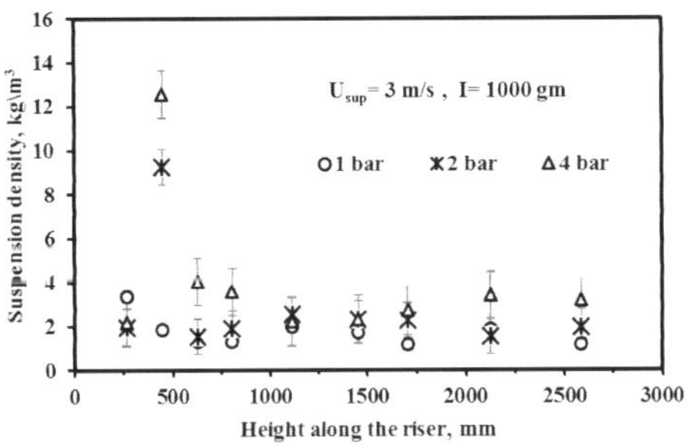

Figure 2. Variation of suspension density with operating pressure along the riser height

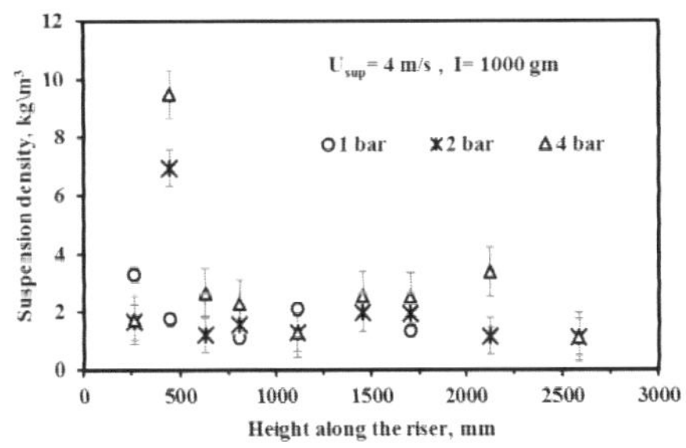

Figure 3. Variation of suspension density with operating pressure along the riser height

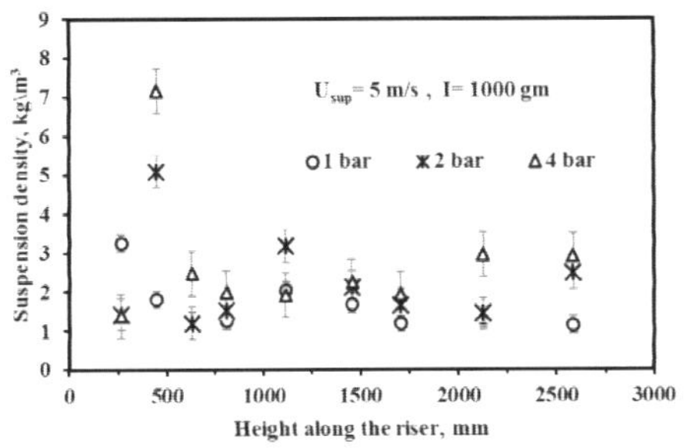

Figure 4. Variation of suspension density with operating pressure along the riser height

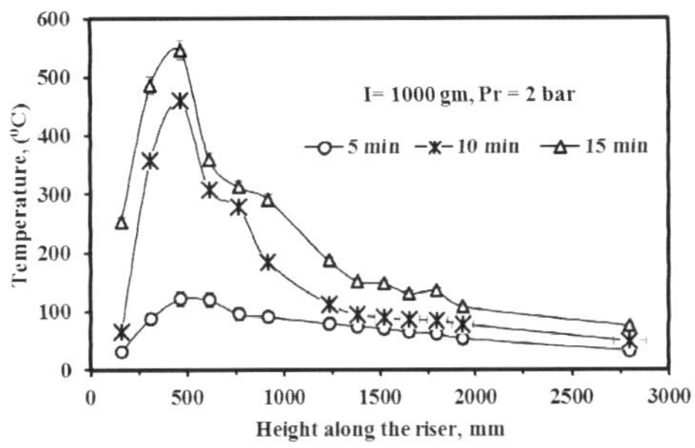

Figure 5. Variation of temperature with time along the riser height

3.2. *Effect of temperature with saw dust as inventory*

The temperature along the height of the riser for operating pressure of 2 bar at different time instances is shown in Fig. 5. The temperature increases in the lower part of the riser i.e. till 500 mm from distributor plate and after that it decreases gradually. With respect to time, the temperature is found to be increasing up to 15 minutes of operating duration, and then starts decaying. The temperature increases rapidly due to the minimum heat loss to the atmosphere at the pressurized condition. During the time when the lower part of the riser achieves a temperature range of 300-550°C, syngas is produced and collected. It takes normally 15 minutes time for gasification of 1000 gm sawdust and after 15 minutes the temperature inside the riser decreases and the syngas is produced up to 20 minutes. Generally, the syngas is collected up to 10 minutes with an inventory of 1000 gm of sawdust.

3.3. *Gasification studies on pressure conditions*

The effect of gas composition with varying pressure from 1 to 4 bar is shown in Fig. 6. The effect of pressure on gas composition is very less significant for hydrogen. With increase in pressure from 1 to 4 bar, hydrogen content decreases from 8.3 to 7.95 %. The percentage of CO increases form pressure 1 to 2 bar and decreases gradually from 2 to 4 bar. There is an increase in CO_2 from 11.2 to 15.2 % with increase in pressure from 1 to 4 bar. The concentartion of methane increases slowly with increase in pressure from 1 to 4 bar. It increases from 3.23 to 4.78 for pressure 1 and 4 bar respectively. The trend of the syngas is different as suggested by different researchers but most of them suggested that there is an increase in CO_2 with increase in gasifying pressure which is also evident in Fig. 6. The reaction $\text{Char} + O_2 \rightarrow CO_2 + CO$ plays an important role in the increase in CO_2 in the syngas composition with an increase in operating pressure.

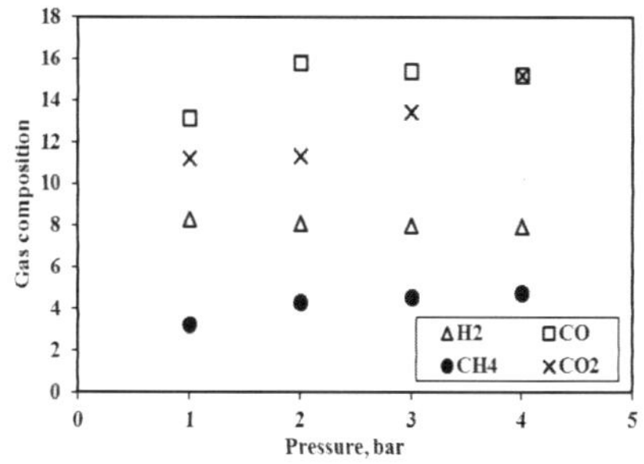

Figure 6. Variation of gas composition with operating pressure

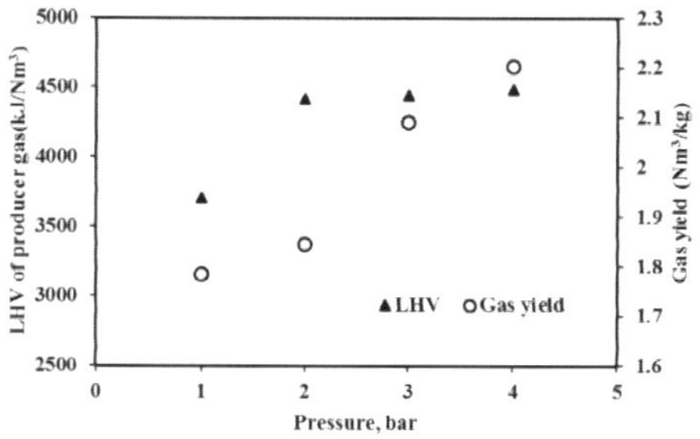

Figure 7. Variation of LHV and gas yield with operating pressure

The lower heating value and gas yield of the syngas with different opearting pressure is represented in Fig. 7. The lower heating value of the syngas is calculated by using the equation $\text{LHV}_{\text{Syngas}} = 4.2 \times \left[(25.7 \times H_2) + (30 \times CO) + (85.4 \times CH_4) + 151.3 \times (C_2H_2 + C_2H_4 + C_2H_6) \right]$. If there is a slight increase in percentage of methane, it affects the value of LHV. The LHV of syngas gas increases with increase in operating pressure, as it is the function of the coefficient of gas compositions. For calculation of LHV, the higher order terms of C_mH_n is neglected, because during gas composation, these values can not be meassured in gas chromotography. There is a impulsive increase in LHV from 1 to 2 bar and then it increases steadily from 2 to 4 bar. The drygas yield increases with an increases in operating pressure. It increases from 1.78 to 2.20 with increases in pressure from 1 to 4 bar. With increase in pressure, N_2 decreases and this is the reason for increase in dry gas yield of the syngas. Moreover, as the temperature increases with pressure, the gas yield also increases with pressure.

The carbon conversion efficiency (CCE) and cold gas efficiency (CGE) have been calculated by using the relationship given by Mallick et al. [15]. The (CCE) of the gasifier increases with an increases in operating pressure (Fig. 8). With increases in pressure, more amount of air is introduced into the riser and with increases in air the amount of O_2 increases. With increase in O_2, there is an increase in gasifying temperature and heat transfer in the riser which results in enhancing the gasification rate and thereby increasing the total carbon content of the syngas. The CCE of the gasifier increases from 50.33 to 79.35 with an increases in pressure from 1 to 4 bar. Similarally, the cold gas efficiency also increases from 37.71 to 59.66 with an increase in pressure from 1 to 4 bar. For the gasification study, the optimal operating condition of pressure = 4 bar where the lower heating value (4746.86 kJ/Nm^3) is higher. The value of CCE and CGE at 4 bar is 79.35 and 59.66 %, respectively.

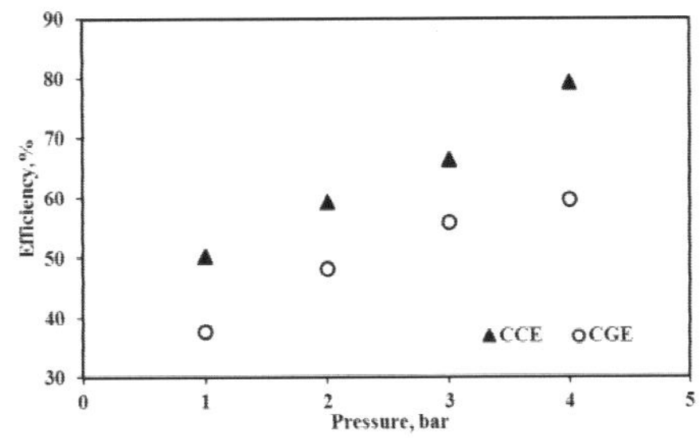

Figure 8. Variation of carbon conversion and cold gas efficiency with operating pressure

4. Conclusions

In this present investigation, hydrodynamic behaviour and gasification study have been carried out in a PCFB gasifier experimentally. Sawdust is considered as a solid bed inventory and the pressure is maintained in the range of 1 to 4 bar. The key findings of the study are summarized below:

- The suspension density increases with operating pressure but decreases with increase in superficial velocity.
- In the syngas, concentration of CH_4 and CO_2 increases with increase in operating pressure.
- The effect of pressure on gas concentration is very less significant for hydrogen. There is a small decreases in hydrogen content from 8.3 to 7.95 % when the pressure increases from 1 to 4 bar.
- The concentration of CO increases when the pressure changes from 1 to 2 bar and decreases gradually from 2 to 4 bar.
- The LHV, gas yield, CCE and CGE has a positive response to the rise in bed operating pressure and are found maximum at 4 bar. The CCE and CGE of the PCFB gasifier are increased by 57 % and 58 %, respectively when the operating pressure is changed from 1 to 4 bar.

References

[1] Saxena R.C., Adhikari, D.K and Goyal, H.B., 2009 Biomass-based energy fuel through biochemical routes: a review. *Renewable and Sustainable Energy Reviews*, **13** (1), pp.167-178
[2] Mallick, D., Mahanta, P. and Moholkar, V.S., 2017 Co-gasification of coal and biomass blends: Chemistry and Engineering. *Fuel*, **204**, pp.106-128
[3] Kumabe, K., Hanaoka, T., Fujimoto, S., Minowa, T. and Sakanishi, K., 2007 Co-gasification of woody biomass and coal with air and steam. *Fuel*, **86** (5-6), pp.684-689
[4] Kurkela, E., Kurkela, M. and Hiltunen, I., 2014 The effects of wood particle size and different process variables on the performance of steam-oxygen blown circulating fluidized-bed gasifier. *Environmental Progress & Sustainable Energy*, **33** (3), pp.681-687
[5] Buragohain, B., Mahanta, P. and Moholkar, V.S., 2010 Biomass gasification for decentralized power generation: The Indian perspective. *Renewable and Sustainable Energy Reviews*, **14** (1), pp.73-92
[6] Chutichai, B., Patcharavorachot, Y., Assabumrungrat, S. and Arpornwichanop, A., 2015 Parametric analysis of a circulating fluidized bed biomass gasifier for hydrogen production. *Energy*, **82**, pp.406-413

[7] Mirmoshtaghi, G., Skvaril, J., Campana, P.E., Li, H., Thorin, E. and Dahlquist, E., 2016 The influence of different parameters on biomass gasification in circulating fluidized bed gasifiers. *Energy Conversion and Management*, **126**, pp.110-123
[8] Xu, G. and Gao, S., 2003 Necessary parameters for specifying the hydrodynamics of circulating fluidized bed risers—a review and reiteration. *Powder Technology*, **137** (1-2), pp.63-76
[9] Kalita, P., Saha, U.K. and Mahanta, P., 2013 Effect of biomass blending on hydrodynamics and heat transfer behavior in a pressurized circulating fluidized bed unit. *International Journal of Heat and Mass Transfer*, **60**, pp.531-541
[10] Datta, S., Sarkar, P., Chavan, P.D., Saha, S., Sahu, G., Sinha, A.K. and Saxena, V.K., 2015 Agglomeration behaviour of high ash Indian coals in fluidized bed gasification pilot plant. *Applied Thermal Engineering*, **86**, pp.222-228
[11] Kumari, G. and Vairakannu, P., 2018 CO2-air based two stage gasification of low ash and high ash Indian coals in the context of underground coal gasification. *Energy*, **143**, pp.822-832
[12] Kumar, K.V., Bharath, M., Raghavan, V., Prasad, B.V.S.S.S., Chakravarthy, S.R. and Sundararajan, T., 2017 Gasification of high-ash Indian coal in bubbling fluidized bed using air and steam–An experimental study. *Applied Thermal Engineering*, **116**, pp.372-381
[13] Miccio, F., Ruoppolo, G., Kalisz, S., Andersen, L., Morgan, T.J. and Baxter, D., 2012 Combined gasification of coal and biomass in internal circulating fluidized bed. *Fuel Processing Technology,* **95**, pp.45-54
[14] Li, K., Zhang, R. and Bi, J., 2010 Experimental study on syngas production by co-gasification of coal and biomass in a fluidized bed. *International Journal of Hydrogen Energy*, **35** (7), pp.2722-2726
[15] Mallick, D., Mahanta, P. and Moholkar, V.S., 2019 Co–gasification of Coal/Biomass Blends in 50 kWe Circulating Fluidized Bed Gasifier. *Journal of the Energy Institute,* https://doi.org/10.1016/j.joei.2019.04.005
[16] Basu, P., 2009 Combustion and gasification in fluidized beds. CRC press.
[17] Motta, I.L., Miranda, N.T., Maciel Filho, R. and Maciel, M.R.W., 2018 Biomass gasification in fluidized beds: A review of biomass moisture content and operating pressure effects. *Renewable and Sustainable Energy Reviews*, **94**, pp.998-1023
[18] Gül, S., Akgün, F., Aydar, E. and Ünlü, N., 2018 Pressurized gasification of lignite in a pilot scale bubbling fluidized bed reactor with air, oxygen, steam and CO_2 agents. *Applied Thermal Engineering*, **130**, pp.203-210
[19] Huang, J., Fang, Y., Chen, H. and Wang, Y., 2003 Coal gasification characteristic in a pressurized fluidized bed. *Energy & Fuels*, **17** (6), pp.1474-1479
[20] Srinivas, T., Gupta, A.V.S.S.K.S. and Reddy, B.V., 2009 Thermodynamic equilibrium model and exergy analysis of a biomass gasifier. *Journal of Energy Resources Technology*, **131** (3), p.031801
[21] Kitzler, H., Pfeifer, C. and Hofbauer, H., 2011 Pressurized gasification of woody biomass—variation of parameter. *Fuel Processing Technology*, **92** (5), pp.908-914
[22] Long H, Qinhui W, Zhongyang L, Nai R, and Guangyi D., 2013 H_2 rich gas production via pressurized fluidized bed gasification of sawdust with in situ CO_2 capture. *Applied Energy*, **109**, pp.36-43
[23] Berrueco, C., Montané, D., Güell, B.M. and Del Alamo, G., 2014 Effect of temperature and dolomite on tar formation during gasification of torrefied biomass in a pressurized fluidized bed. *Energy*, **66**, pp.849-859
[24] Sánchez, C., Arenas, E., Chejne, F., Londoño, C.A., Cisneros, S. and Quintana, J.C., 2016 A new model for coal gasification on pressurized bubbling fluidized bed gasifiers. *Energy Conversion and Management*, **126**, pp.717-723.
[25] Doebelin, E.O. and Manik, D.N., 2007 Measurement systems: application and design. McGraw Hill Education; 6 edition.

IOP Conf. Series: Materials Science and Engineering **605** (2019) 012004 doi:10.1088/1757-899X/605/1/012004

Demand Response Management in the Presence of Renewable Energy Sources using Stackelberg Game Theory

Pushpendra Singh[1], Akash Talwariya[2] and Mohan Kolhe[3]

[1] Associate Professor, JK Lakshmipat University, Jaipur, India
[2] Research Scholar, JECRC University, Jaipur, India
[3] Professor, University of Agder, Norway
akash.talwariya@gmail.com

Abstract. A demand response algorithm for electricity trading using renewable energy sources using stackelberg game methodology. Energy consumption is rapidly increasing day by day and to maintain the balance between energy consumption and generation needs solution to generate electricity from renewable energy sources and demand response methodology is essential. Demand response (DR) algorithm maintain the balance and reduce the peak to average energy consumption ratio. Large number of constraints (i.e. number of appliance at consumer end, different consumers, renewable energy sources, electric vehicles, etc.) interface make the DR algorithm complex. Stackelberg game theory is provided the solution based on one leader N follower strategy where leader first decide their best response and followers select their best response on the basis of leaders strategy and then leader again optimize the new strategy. Process is repeat again and again to find nash equilibrium between consumers and utilities. Proposed methodology increase the generation from renewable energy sources and manage the appliances. Simulation results showed that stackelberg game with demand response algorithm is effective for optimal energy consumption control of appliances.

Keywords- Stackelberg game theory; energy management center; renewable energy sources; electricity trading.

1. Introduction

Energy consumption pattern varies with respect to time, energy demand is higher during peak hours and low during normal hours. Demand response is providing promising solution for energy management [1-2]. Incentive based mechanism attracting consumers for DR. It convinces consumers to consume less energy during peak hours and manage their energy demand from peak hours to normal hours [3]. A large amount of energy is generated through fossil fuels they harm the environment drastically [4-5]. Renewable energy sources (RES) near load centers provide probable solution and reduce the dependency of consumers on fossil fuels. Energy generation through RES are random in character and reliability of generation is very low. Introduction of electric vehicles (EV's) increase the reliability of the renewable energy sources by providing availability of electricity during off generation period [6-7]. For long term investigation real time pricing is expected to provide better results. An Energy management center (EMC) is established between utility and consumers. Prime objective of the EMC is to minimize the dependency of consumers on grid and provide maximum amount of energy through RES [8-10]. By the introduction of RES, EV's, DR, real time pricing make the system complex and need the solution for optimization [11]. Game theory providing the solution for optimization demand response.

Game theory is mathematical tool based on desired rules to find the win-win situation between all the agents. Different agents compete to maximize their benefits. Demand response algorithm is classified agents in utility of consumers [12-13]. On the basis of different agents participated game theory is classified in two categories, the game played between different consumers and providing nash equilibrium between consumers e.g. day ahead price, real time price, incentive based games etc. Second game is played between consumer and utility to provide benefits to utilities and on the basis of their strategy consumer updated their strategy to find nash equilibrium is classified as stackelberg game [14-15].

Stackelberg game is played between consumer and utility, utility is participated as leader and they first select the best strategy amongst their strategy set. Consumers participated as followers and they select their best strategy based on utility best response and again by considering the response of followers leader update their response. This process repeat again and again till the both leader and follower get their combined best response. Best response of utility and consumer is achieved at a particular condition is known as nash equilibrium where consumer as well as utility both are satisfied. Main Contribution of this paper are as follows.

1. To reduce the dependency on fossil fuels generate electricity through renewable energy sources near load centers.
2. By introduction of electric vehicles with RES increase the reliability of the system.
3. A one leader N- follower methodology for demand response is proposed using stackelberg game for residential consumers.
4. A dissatisfaction function is defined for different appliances and a stackelberg equilibrium is obtained which define the coalitions of different players.

Rest of the paper is organized as follow. Section 2 introduced the stackelberg model in demand response methodology with mathematical modeling include the energy consumption model, utility model and renewable energy generation model. Section 3 elaborate the process of stackelberg game and nash equilibrium of stackelberg game. Simulation results of residential consumers equipped with smart appliance is discussed in section 4 and section 5 conclude the paper with research finding and outcomes.

2. Stackelberg Game Model

Stackelberg game is played between consumers and utility where utility is participated as leader and consumers are participated as followers [16-17]. Some assumptions are taken as following:

1. All active consumers equipped with Energy management center for their energy management.
2. Each consumer having two category of appliances first in interruptible appliances (i.e. electric vehicles, dishwasher, washing machine, AC's etc.) and second are uninterruptible appliances (i.e. lightning, refrigerators etc.).
3. Energy management center is equipped with renewable energy sources as primary energy supplier for consumers.

2.1 Consumer Model

The energy management center receives the real time energy price few minutes ahead e.g. 15 to 20 minutes [18-19], and manage the energy consumption of all connected appliances as per the response received from utility.

In stackelberg game model shown in figure 1 utility with the help of EMC behave as a leader upload the strategy of energy price to the consumers, on the basis of energy consumption each and every consumer define its strategy of consumption and after taking in consideration of consumer best strategy utility update their own strategy. The process run again and again to find nash equilibrium between consumer and utility.

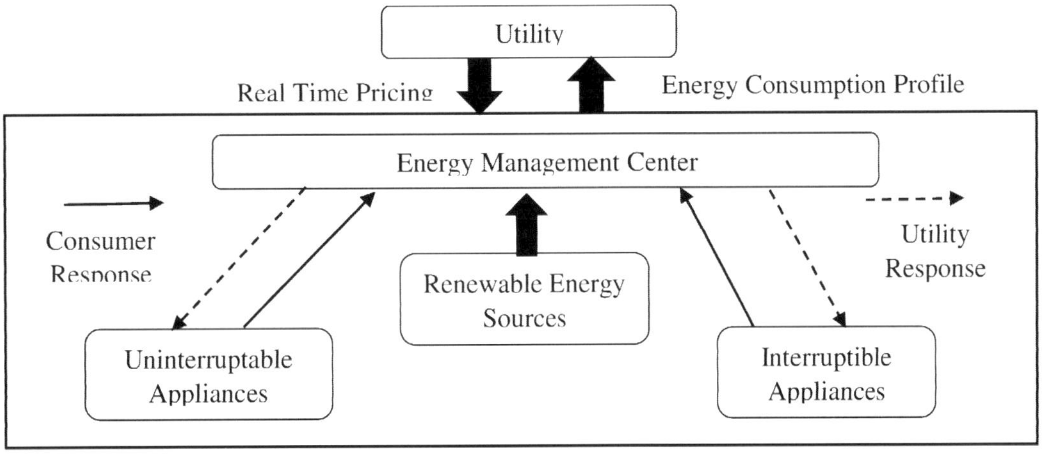

Figure 1. Energy trading process

2.1.1. Regulatory cost

Utilities are investing on infrastructure, generating, transmission and distribution of energy to provide reliable energy to consumers. Utilities charged in terms of tariff for recovery of their investments. Energy management centers for providing benefits to consumers providing best suitable strategy of consumption. Consumer payoffs are based on two factors first is constraints wattage and second is probability of the constraint to maximize the payoff of the consumer. Conditional probability of $m_i \in M_i$ is $p(m_i|m_{-i})$ define in (1) where $m_{-i} = (m_1, \ldots, m_{i-1}, m_{i+1} \ldots M_i)$ is the combination of all consumers except i^{th} consumer.

$$p(m_i|m_{-i}) = \frac{p(m_{-i,m_i})}{\sum p(m_i)} \tag{1}$$

Payoff function of consumer i for category m_i can be expressed as EP*i* (m*i*) in (2):

$$EPi\ (mi) = \sum_{m_i \in M_i} (p(m_i|m_{-i}, x_c)x_c(m_i))p(m_i|m_{-i}) \tag{2}$$

Where, $x_c(m_i)$ denotes the wattage of constraint c for category m_i.

2.1.2. Dissatisfaction cost function

Dissatisfaction cost function denoted the consumer satisfaction at a particular time to use the appliances based on total energy consumption. Dissatisfaction cost function is varied from positive to negative by comparing with median of total energy consumption denoted as K_j. If energy consumption at a particular time is less than median energy consumption, dissatisfaction cost function denoted positive value. Positive dissatisfaction function meaning is that the current strategy is dissatisfied to consumers. If the energy consumption at a particular time is higher as compared to median energy sources, dissatisfaction function is denoted negative and the meaning is that consumers are satisfied with the strategy provided by EMC based on DR algorithm. Dissatisfaction cost function is calculated by (3).

$$\Phi_j(x_j) = e^{\Upsilon_j\left(1-\frac{x_j}{k_j}\right)} - 1,\ \Upsilon_j > 0 \tag{3}$$

In (3), Υ_j is denoted as priority factor of appliances j ($\forall j \in J$) based on their utilization. Value of Υ_j is higher for higher priority and lower for lower priority of appliances.To minimize the energy consumption, the utility function of an appliance is denoted as $U_j(c, x_j)$ have the summation of regulatory cost function and dissatisfaction cost function in (4).

$$U_j(c, x_j) = EPi\ (mi) + \omega\ .\ \Phi_j(x_j),\ \omega > 0 \tag{4}$$

In (4), ω is defined as the satisfaction over the given period. Energy consumption period is divided in three different durations e.g. peak energy consumption duration, modest energy consumption duration and least energy consumption duration. The value of ω is higher indicates that consumer is concern about reducing dissatisfaction. In energy consumption cost model, need to obtain an optimal strategy x_j foe each appliance with respect to defined strategy of utility. The appliance aims to minimize the cost, the problem is formulated as in (5):

$$min \quad U_j(c\ ,\ x_j) \tag{5}$$

$$s.t.\ x_j^{min} \leq x_j \leq x_j^{max} \quad \forall j \in J \tag{6}$$

2.1.3. Renewable Generation Cost

RES generation random characteristics make it complex to design the generation from renewable energy sources. RES are the main source of energy and consider satisfaction function to supply energy to utility company [20]. Renewable generation cost model is denoted by U_R in (7):

$$U_R = [R_R + C_R].X_R \tag{7}$$

Where, R_R denoted as the revenue invested on energy generation through RES in EMC and C_R is denoted as the cost of energy brought by utility and X_R is the satisfaction parameter of RES generation.

2.2. Utility Model

The objective of utility is to get maximum benefit by charging suitable tariff from consumers. Conventional energy tariff models are particularly based on energy consumption all the time. Proposed model with stackelberg game provide an opportunity to the utilities to update their strategy based on consumers strategy. The objective of the consumer is to minimize the dissatisfaction cost function by updating their own strategy and reduce the tariff cost by adding low cost generation from renewable energy sources is defined in (8).

$$U_{utility} = \sum_{i=1}^{I} EPi\ (mi)\ + U_R - \sum_{j=1}^{J} \omega\ .\ \Phi j(xj) \tag{8}$$

Where, $EPi\ (mi)$ denotes the gross benefit gained by consumer by managing the demand of each appliances and U_R denotes as the energy brought cost from RES. $\sum_{j=1}^{J} \omega\ .\ \Phi j(xj)$ is dissatisfaction function will reduce overall cast with increasing satisfactory factor.

3. Stackelberg Game Analysis

SNE is obtained by maximize the utility strategy based on best response of consumers strategy [3-5]. Energy consumers are able to minimize their energy consumption by applying best strategy on the basis of utilities strategy and utilities update their strategy on the basis of consumers strategy is called SNE and all categorized consumers and utility are agreed on coalitions. The SNE is defined as (c^*, x^*) for the best strateghy of consumers $x^* = [x_1^*, x_2^*, ..., x_N^*]$ and c^* denotes the best response of utility are define by (9-10):

$$(c^*, x^*) = max_{(c,x)}\, U_{utility}\,(c, x^*) \tag{9}$$

$$s.t.\ min\ min_x\ U_j\,(c,x_j) \quad \forall j \in J \tag{10}$$

4. Simulation Results:

Household consumers are equipped with interruptible and uninterruptable appliances, interruptible appliances are one washing machines denoted as WM, one dish washers denoted as DW, one air conditioners denoted as AC and one electric vehicle denoted as EV. Uninterruptible appliances are two lights denoted as L1 and L2, and one refrigerators RF. Appliance characteristics are listed in Table-I, each appliance have significant value of Υ_j denotes the probability of operation [21-22].

Table 1. Appliance Characteristics

Appliance Type	**Interruptible Appliances**				**Uninterruptible Appliances**		
Appliance	**AC**	**EV**	**DW**	**WM**	**L1**	**L2**	**RF**
Υ_j	4.5	3.5	5	4	2	2.5	1.5
Power Rating (kWh)	1.5-3	1.5-3	2	1-3	0.5-1.5	0.5-1.5	1.5-3
Duration of Operation (hr.)	8-10	2-4	2	2	8-10	8-10	24

Appliance probability is defined that RF, L1 and L2 have lowest priority of operation means they operate on priority. A satisfaction factor ω is denoted by dividing a day in three different energy consumption durations listed in Table-2. Satisfaction factor is higher in off-peak period and less in peak period where utility expect less energy consumption from each appliances [7].

Table 2. Satisfied Weight Factor for Different Time Durations

Duration	**Off-peak duration (1:00 am to 7:00 am & 12:00 pm to 1:00 am)**	**Mid-peak duration (10:00 am to 5:00 pm & 11:00 pm to 12:00 pm)**	**Peak duration (8:00 am to 10:00 am & 6:00 pm to 10:00 pm)**
ω	0.18	0.1	0.07

For the real time pricing, we adopted the on-line data of energy consumption price from ComEd (Commonwealth Edison Company) for the date may 11, 2019 [23-24]. Figure 2 represent the energy price of appliances as per the priority and provide stackelberg nash equilibrium between consumer and utility for interruptible and uninterruptible appliances respectively in (a) & (b) prices are higher for interruptible appliances as their energy consumption priority is higher.

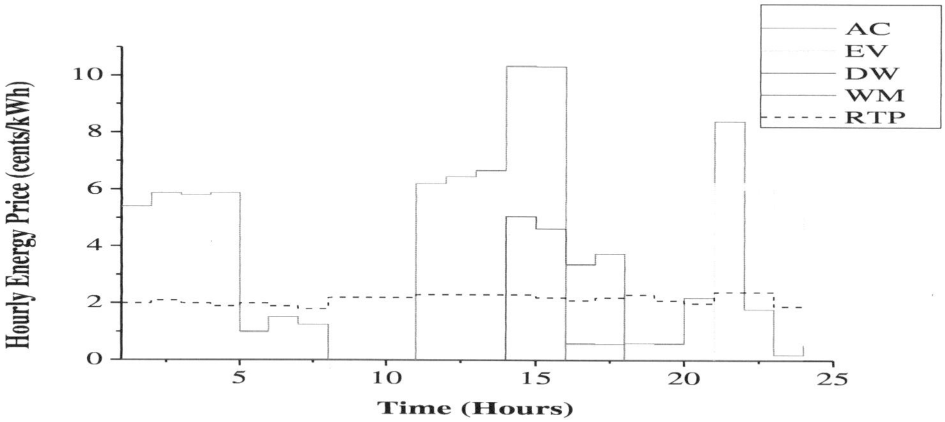

(a)

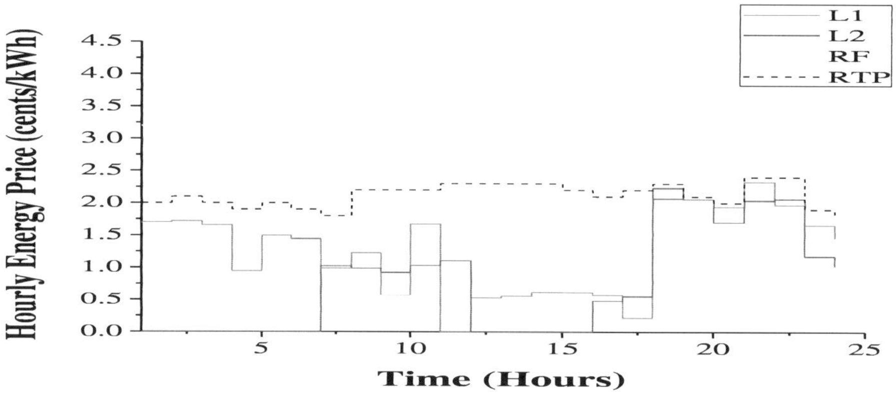

(b)

Figure 2. Hourly Energy Consumption for Residential Consumers (a) Interruptible Appliances (b) Uninterruptible Appliances

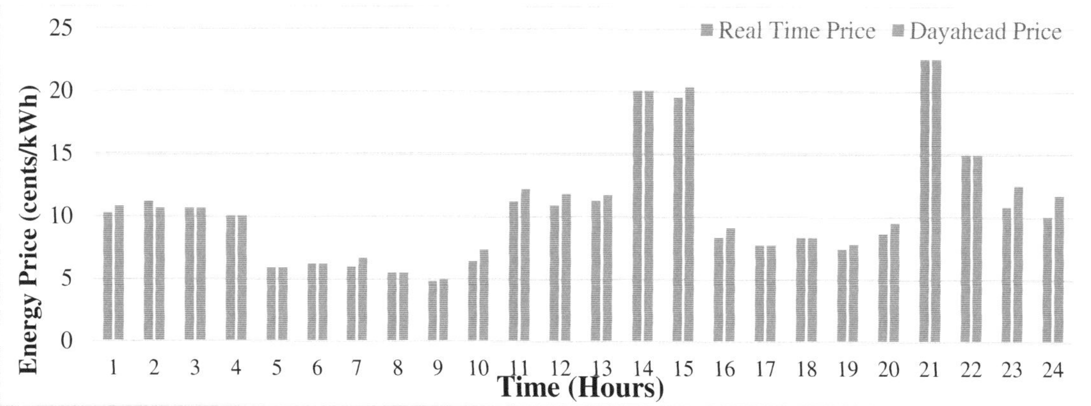

Figure 3. Comparative Analysis of Real Time Price and Day Ahead Price

Real time price model is compared to day ahead energy price model produced by same ComEd company for the date may 11, 2019. Figure 4 shows the comparison between RTP and day ahead price for the proposed stackelberg nash equilibrium for residential consumers. Results shows that total energy price reduced from 259.35 ¢ to 249.91 ¢ when using RTP instead of day ahead price. Average hourly price reduced from 10.80 ¢/hour to 10.37 ¢/hour.

5. Conclusion

A novel real time price based demand response approach using stackelberg game theory is used for retail electricity market with energy management center to control the consumption from renewable energy sources. Proposed stackelberg game based demand response model is able to manage the energy consumption for less energy consumption during high price range and higher during low consumption range and providing nash equilibrium where utility and consumer both get benefitted. Energy consumption patterns are also compared with day ahead energy consumption and providing better results. RES are connected at consumer manage the RTP and maintain lesser compared to day ahead price. In future the practically investigated the probability of appliances on existing smart grids.

6. References

[1] Siano, Pierluigi. "*Demand response and smart grids—A survey.*" Renewable and sustainable energy reviews 30 (2014): 461-478.
[2] Ghalelou, Afshin Najafi, et al. "A stochastic self-scheduling program for compressed air energy storage (CAES) of renewable energy sources (RESs) based on a demand response mechanism." *Energy conversion and management*120 (2016): 388-396.
[3] Ghazvini, Mohammad Ali Fotouhi, et al. "A multi-objective model for scheduling of short-term incentive-based demand response programs offered by electricity retailers." *Applied energy* 151 (2015): 102-118.
[4] De Matos, José G., Felipe SF e Silva, and Luiz A. de S. Ribeiro. "Power control in ac isolated microgrids with renewable energy sources and energy storage systems." *IEEE Transactions on Industrial Electronics* 62.6 (2015): 3490-3498.
[5] Meibom, Bent Sorensen, Peter. "A global renewable energy scenario." *International Journal of Global Energy Issues* 13.1-3 (2000): 196-276.
[6] Singh, Pushpendra, D. P. Kothari, and Mool Singh. "Feasibility of interconnected distribution network for the integration of distributed energy resources using ga approach." *International Review of Electrical Engineering* (2013).
[7] Singh, Pushpendra, D. P. Kothari, and Mool Singh. "Integration/Analysis of Green Energy and Conventional Electricity Generation with Distribution Networks Interconnected." *International Review on Modelling and Simulations* 5.5 (2012).
[8] D. Sharma, A. Talwariya, A. Pandey and P. Singh, "Shrouded problems of solar power plant and recommendations," *2017 Innovations in Power and Advanced Computing Technologies (i-PACT)*, Vellore, 2017, pp. 1-5. doi: 10.1109/IPACT.2017.8245160
[9] Kaushik, Rini. "Energy management costs for a data center." U.S. Patent Application No. 10/037,501.
[10] Heller, Brandon, et al. "Elastictree: Saving energy in data center networks." *Nsdi*. Vol. 10. 2010.
[11] Talwariya, Akash, and Pushpendra Singh. "DEMAND SIDE MANAGEMENT: A SURVEY OF SPT & TOU." *International Journal of Creative Research Thoughts (IJCRT)* 6.1 (2018): 1675-1678.
[12] A. Talwariya, D. Sharma, A. K. Pandey and P. Singh, "An execution of smart grid with game theory," *2016 International Conference on Recent Advances and Innovations in Engineering (ICRAIE)*, Jaipur, 2016, pp. 1-4. doi: 10.1109/ICRAIE.2016.7939503
[13] P. Yang, G. Tang, and A. Nehorai, "A game-theoretic approach for optimal time-of-use electricity pricing," *IEEE Trans. Power Syst.*, vol. 28, no. 2, pp. 884–892, May 2013.
[14] A. Talwariya, P. Singh, M. Kolhe and S. K. Sharma, "A Game Theory Approach and Tariff Strategy for Demand Side Management," *2018 3rd International Conference and Workshops on*

Recent Advances and Innovations in Engineering (ICRAIE), Jaipur, India, 2018, pp. 1-5. doi: 10.1109/ICRAIE.2018.8710409.
[15] Tushar, Wayes, et al. "Price discrimination for energy trading in smart grid: A game theoretic approach." *IEEE Transactions on Smart Grid* 8.4 (2016): 1790-1801.
[16] Yu, Mengmeng, and Seung Ho Hong. "A real-time demand-response algorithm for smart grids: A stackelberg game approach." *IEEE Transactions on smart grid* 7.2 (2015): 879-888.
[17] Zhou, Zhenyu, et al. "Game-theoretical energy management for energy Internet with big data-based renewable power forecasting." *IEEE Access* 5 (2017): 5731-5746.
[18] F.-L. Meng and X.-J. Zeng, "A Stackelberg game-theoretic approach to optimal real-time pricing for the smart grid," *Soft Comput.*, vol. 17, no. 12, pp. 2365–2380, 2013.
[19] S. Maharjan, Z. Quanyan, Z. Yan, S. Gjessing, and T. Basar, "Dependable demand response management in the smart grid: A Stackelberg game approach," *IEEE Trans. Smart Grid*, vol. 4, no. 1, pp. 120–132, Mar. 2013.
[20] Talwariya, Akash, Pushpendra Singh, and Mohan Kolhe. "A stepwise power tariff model with game theory based on Monte-Carlo simulation and its applications for household, agricultural, commercial and industrial consumers." *International Journal of Electrical Power & Energy Systems*111 (2019): 14-24.
[21] Han, Jinsoo, et al. "Smart home energy management system including renewable energy based on ZigBee and PLC." *2014 IEEE International Conference on Consumer Electronics (ICCE)*. IEEE, 2014.
[22] Kumar, Sasi, et al. "Smart Home Energy Management System Including Renewable Energy Based on Zigbee and ARM9 Microcontroller." (2017).
[23] Yu, Rong, et al. "Balancing power demand through EV mobility in vehicle-to-grid mobile energy networks." *IEEE Transactions on Industrial Informatics* 12.1 (2015): 79-90.
[24] Tushar, Wayes, et al. "Cost minimization of charging stations with photovoltaics: An approach with EV classification." *IEEE Transactions on Intelligent Transportation Systems* 17.1 (2015): 156-169.

Fault Diagnosis of On-load Tap-changer Based on Vibration Signal

Wenquan Huang, Changqing Peng, Liang Jin, Jiacheng Liu, Yuqi Cai

Huaqiao University, Xiamen 361021 P.R. China.

peter81015 @hqu.edu.cn

Abstract. On-load tap-changer (OLTC) is an important part of large-scale load transformers, and its fault will directly affect the reactive power flow regulation of the power system and the stability of the load center voltage. In this paper, a fault state diagnosis method based on vibration signal is proposed for fault diagnosis of tap changer. The vibration signals are decomposed by wavelet packet transform and calculated energy entropy to obtain the state information of the tap changer. The experimental data shows that the state characteristics are obvious under different faults, which lays a foundation for in-depth study of fault diagnosis.

1. Introduction

The transformer is one of the core equipment in the power system, and the on-load tap-changer (OLTC) is the only operating component of the power transformer. It plays an important role in regulating the reactive power flow and stabilizing the load center voltage. Once a tap changer fails, it will directly affect the stability and safety of the transformer and power system.

Vibration occurs during the OLTC action process. Different states of the switch have different effects on the vibration, that is, the state information of the OLTC is included in the vibration signals. Some scholars have proposed to use the vibration to diagnose the tap changer fault. Kang et al[1-3] used wavelet analysis to extract the time-domain features such as amplitude, time interval and wavelet ridge obtained from the envelope, and compared the normal and fault characteristics by self-organizing mapping method. Rivas et al.[4-6] used the wavelet transform to decompose the vibration signals. According to the information of the peak time and amplitude of the vibration signal, the characteristics of the normal and typical faults of the OLTC model were compared, and the evaluation criteria of the typical working state of the OLTC were obtained. Zhao Tong et al.[7] reconstructed the original signal in high-dimensional space according to the chaotic characteristics of the vibration signal during the OLTC switching process, and defined the phase point spatial distribution coefficient to identify the normal and fault state of OLTC. Zhang Huifeng[8] proposed using the Hilbert-Huang transform algorithm to analyze the vibration signal in the time-frequency domain, and combined with the frequency change to diagnose the typical fault of OLTC.

In this paper, the vibration sensor is installed at the appropriate position of the OLTC to obtain the vibration signal, and the signals are analyzed to obtain the characteristic information from the aspects of timing, spectrum and energy entropy. The experimental results show that the state characteristics are obvious under different faults, which provides a condition for further fault diagnosis.

2. Signal acquisition and preprocessing

The accelerometer is mounted on the OLTC end cap, and the signal is transmitted to the data acquisition device via the cable. Finally, the vibration data will be transferred to the PC for storage and processing. Because the experimental site environment and electromagnetic environment are relatively harsh, and most of the vibration signals are weak, it is necessary to analyze and preprocess the vibration signals. The signal acquisition system and preprocessing flowchart is shown in Figure 1.

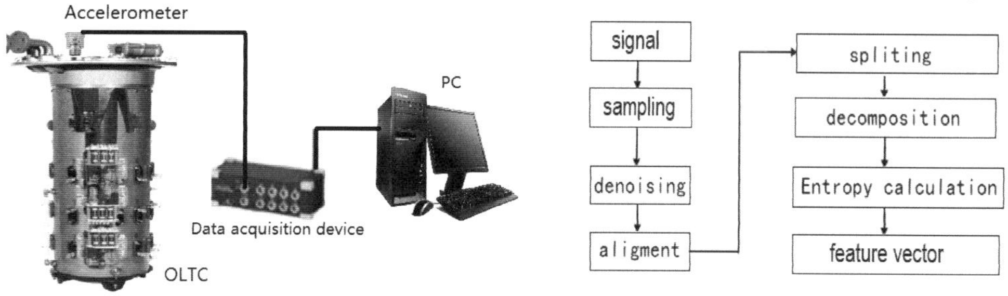

Figure 1. signal acquisition system and preprocessing flowchart

When the signals are sampled synchronously, the vibration sensors are installed at different positions such as the side wall and the top of the on-load tap-changer. After the preliminary analysis and comparison of the signals, the position with stronger vibration signal is selected as the measurement point of this experiment.

As we know that the vibration signal is mainly concentrated in the low frequency 500Hz and below from the spectrum analysis of the signal[8]. The electromagnetic environment of the experimental site is complex, so that it is often necessary to filter the sampled data. In the paper, the wavelet algorithm is used for filtering and noise reduction processing, which can reduce the high-frequency random noise mixed into the vibration signal, while the low-frequency vibration signal information and the characteristic parameters of vibration signal are retained.

In the process of sampling the vibration signal of the on-load tap-changer, the vibration signals are unaligned due to human operation habits and other factors. In order to facilitate subsequent analysis and processing, the vibration signal data needs to be translated and aligned.

Through the principle of the tap-changer and the vibration signal analysis of the normal situation, it is found that the whole vibration signal can be clearly divided into two stages, namely the preparing section and the in-position section, as shown in the following figure. The preparing section is mainly the motor running, driving the mechanical structure to store the spring energy, and the vibration signal is relatively stable at this stage. The in-position section is a segment that the tap-changer works in a predetermined order and the transition resistance is pre-switched to realize the gear shift operation. The vibration signal is abruptly changed, and there are three signal peaks.

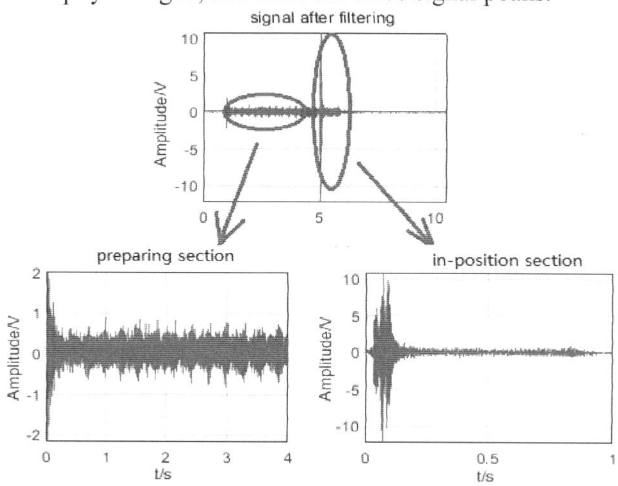

Figure 2. Vibration signal segmentation and signal amplification

Because there are obvious differences between the two stages, it is necessary to divide the whole process into the preparing section and the in-position section, and perform wavelet packet decomposition and calculate energy entropy to extract state features of OLTC.

3. Wavelet packet decomposition and IMF energy entropy extraction

3.1. Signal decomposition and energy entropy calculation introduction

The traditional Fourier transform is suitable for signals that are more stable over a long period of time. The signal after orthogonal decomposition of the wavelet packet has the characteristics of independent signal and energy conservation in each frequency band, and is more suitable for time-frequency analysis and energy spectrum analysis of vibration signals. Therefore, the wavelet packet transform is applied to the fault diagnosis of the tap changer. By selecting the appropriate wavelet basis function, the vibration signal is decomposed into several sub-components by wavelet packet, and then the energy value of each sub-component is calculated. After normalization, the energy entropy vector is constructed as the vibration feature. The process is as follows:

(1)Vibration signals are preprocessed, including filtering, alignment, and segmentation.

(2)The j-layer wavelet packet decomposition is performed on the vibration signal data S. Taking j=4 as an example, $M=2^4$ sub-band signals are obtained on the fourth layer. The wavelet packets coefficients corresponding to the $k\left(k=1,2,\cdots,16\right)$ subband is $d(j,k)$.

(3)Calculate the energy value of each sub-band of the bottom layer, and the energy on the k-th sub-band is:

$$E\left(j,k\right)=\frac{1}{N}\sum_{i=1}^{N}\left|d_i\left(j,k\right)\right|^2 \tag{1}$$

where N is the length of the k-th subband signal.

Then the total energy of the signal is:

$$E=\sum_{i=1}^{M}E_i(j,k) \tag{2}$$

If $\varepsilon_{jk}(i)=E_{jk}/E$, then $\sum_k \varepsilon_{jk}=1$.

(4) Defining the wavelet packet energy entropy is:

$$H_{jk}=-\sum_{i=1}^{N}\varepsilon_{jk}\left(i\right)\log\left|\varepsilon_{jk}\left(i\right)\right| \tag{3}$$

Where H_{jk} is the k-th sub-band wavelet packet energy entropy of the j-layer.

The energy entropy of each wavelet packet of the bottom layer is obtained by equation (3). Taking 4 layers as an example, the energy entropy eigenvector is recorded as:

$$T=\left[H_{41},H_{42},\cdots,H_{415},H_{416}\right] \tag{4}$$

If $H=\sqrt{\sum_{i=1}^{16}\left|H_{4i}\right|^2}$, then the energy entropy feature vector can be normalized to:

$$T^{'}=\left[\frac{H_{4,1}}{H},\frac{H_{4,2}}{H},\cdots,\frac{H_{4,15}}{H},\frac{H_{4,16}}{H}\right] \tag{5}$$

IOP Conf. Series: Materials Science and Engineering **605** (2019) 012005 doi:10.1088/1757-899X/605/1/012005

3.2. Wavelet packet decomposition of the preparing section and the in-position section in the normal state of 5-6 gear

After the vibration signal is filtered and noise-reduced, the wavelet packet decomposition transform described above is used. Select ' db3 ' as the basis function to perform four-layer wavelet packet decomposition on the vibration data. The result is shown in Figure 3.

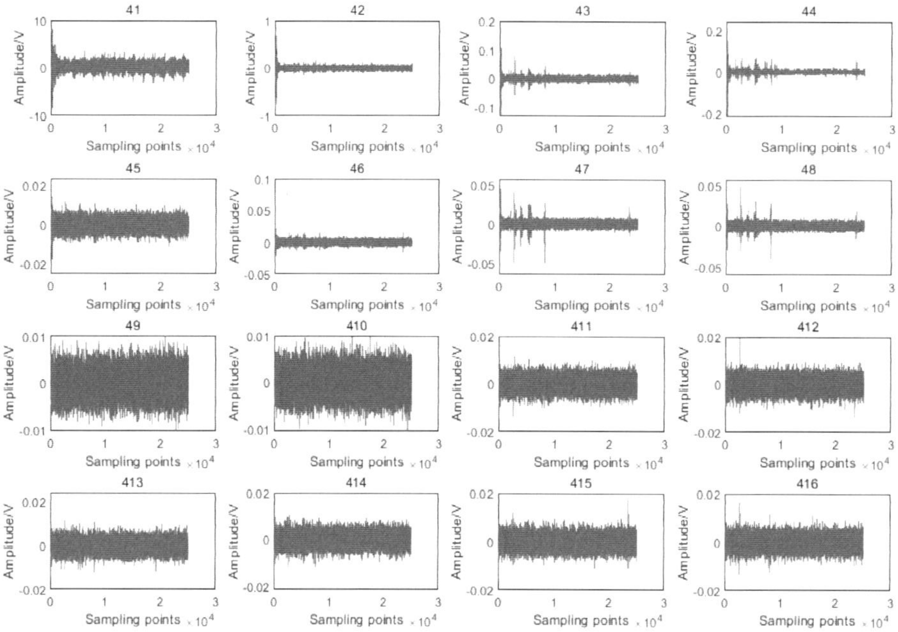

a.decomposition of preparing section

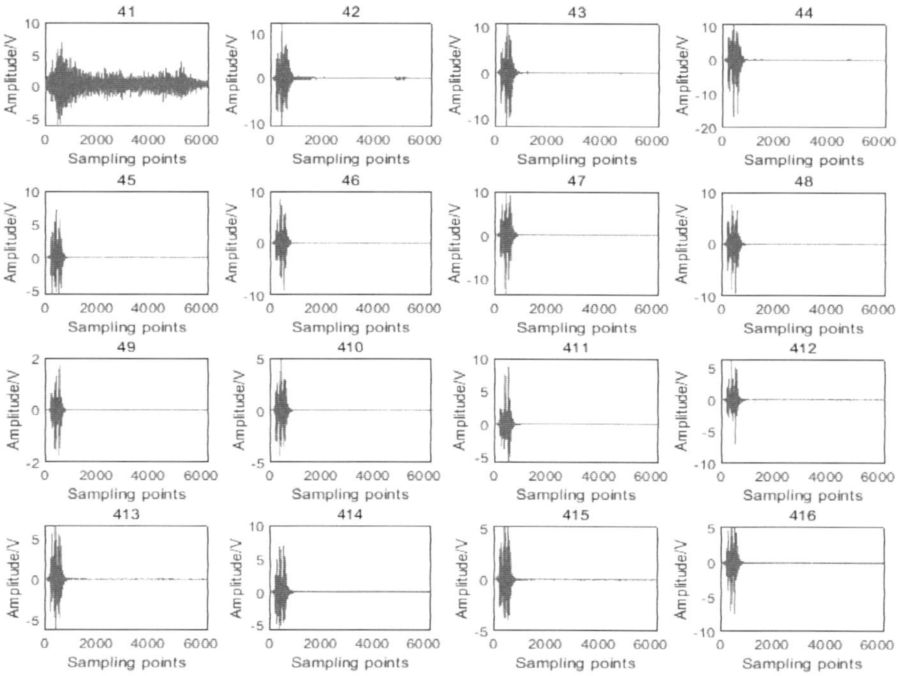

b.decomposition of in-position section

Figure 3. Four-layer wavelet packet decomposition of vibration signal

3.3. Energy entropy calculation and distribution

On the basis of the wavelet packet decomposition, the energy entropy of each component is further calculated and normalized. In the same way, the vibration signals under the fault conditions such as switching-late and spring-breakage, etc. are also processed and decomposed. Then the energy entropy of each sub-band is obtained. The comparison result by the histogram is shown in Figure 4.

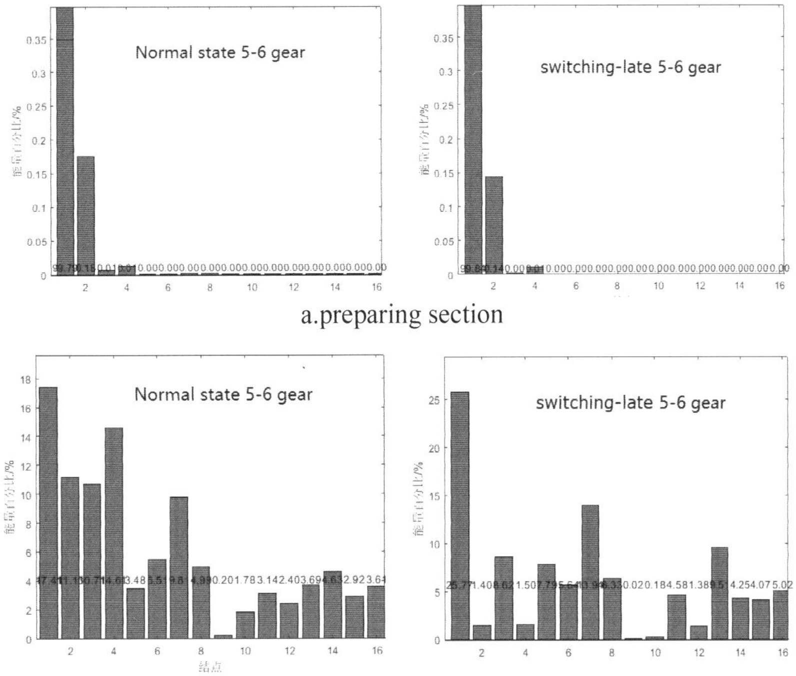

a.preparing section

b.in-position section

Figure 4. Energy entropy distribution of normal state and switching-late state

It can be seen from the comparison of the Figure 4 that the fault occurs mainly in the in-position segment. Comparing the switching-late state with the normal state, the energy entropy distribution in the in-position section are significantly different in Figure 4b. These feature information can be used as the eigenvector of the support vector machine. While the energy entropy distribution of the preparing sections in Figure 4a is basically the same, consistent with the actual physical process.

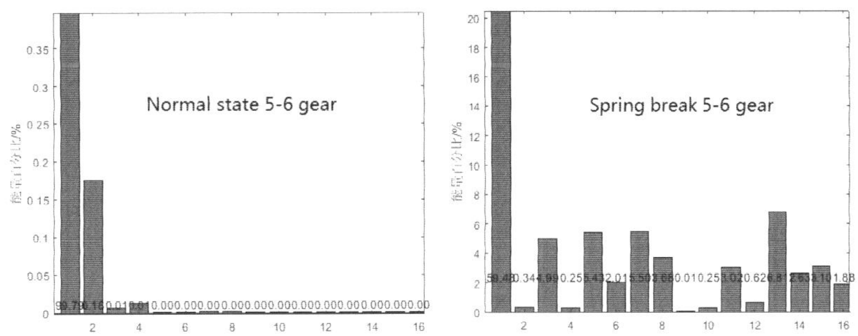

a.preparing section

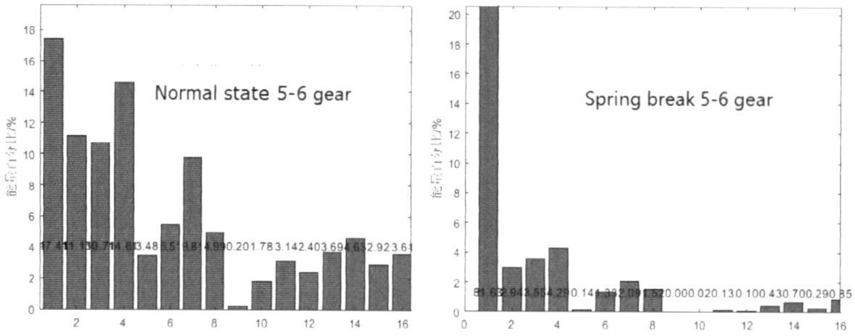

b.in-position section

Figure 5. Energy entropy distribution of normal state and spring-breakage state

It can be seen from the figure that the spring break will affect the normal energy storage of the preparing section and even affect the gear shifting action of the in-position section. In Figure 4a, the energy entropy distribution under the fault of spring-break is obviously different from the normal state. The energy entropy of wavelet component 1 is relatively large, and the other components also account for a considerable proportion. In the normal state, except the wavelet component 1, the rest of the wavelet The entropy of energy is much smaller and can even be ignored. In Figure 4b, the gear shifting action is also affected and the vibration signal waveform is weak due to spring breakage. The energy entropy distribution is different from the others.

4. Conclusions

The research shows that the vibration signal can be used to diagnose the working state of the tap changer. In this paper, the vibration signal of the tap-changer are filtered and divided into the preparing section and an in-position section. Then the vibration signals are decomposed by wavelet packet transform to obtain each vibration signal component, and the energy entropy of each signal components are calculated. The statistical data shows that there are significant differences in the energy entropy distribution of the tap changer in different states (such as switching too late, spring break), which can be used as the basis for discriminating the normal state and the fault state.

In further research, the energy entropy of these vibration signals will be used to construct the feature vector of SVM to verify the feasibility of this method in OLTC fault diagnosis applications. At the same time, the vibration signal of this paper is measured under the light load condition of OLTC, and it is still to be verified under the actual load condition.

Funding

The The authors disclosed receipt of the following financial support for the research, authorship, and/or publication of this article: Natural Science Foundation of Fujian Province, China (Project No. 2019J01054).

References

[1] Kang, P. , & Birtwhistle, D. . (2003). Characterisation of vibration sigsnals using continuous wavelet transform for condition assessment of on-load tap-changers. Mechanical Systems & Signal Processing, 17(3), 561-577.

[2] Kang P, Bitwhistle D. Condition assessment of power transformer on-load tapchangers using wavelet analysis and self-organizing map: field evaluation.IEEE Trans Power Deliv 2003;18:1;

[3] Kang, P., & Birtwhistle, D. (2001). Condition assessment of power transformer on-load tap-changers using wavelet analysis. Power Delivery IEEE Transactions on, 16(3), 394-400.

[4] Rivas E , Burgos J C , Garcia-Prada J C . Condition Assessment of Power OLTC by Vibration Analysis Using Wavelet Transform[J]. IEEE Transactions on Power Delivery, 2009, 24(2):687-694.

[5] Rivas-Trujillo E , Edwin Laniado-Jácome, José Luis San-Román, et al. Characterizing the diverter switch of a load tap changer in a transformer using wavelet and modal analysis[J]. Engineering Structures, 2010, 32(10):3011-3017.

[6] Rivas, E., Burgos, J. C., & Garcia-Prada, J. C. (2010). Vibration analysis using envelope wavelet for detecting faults in the oltc tap selector. IEEE Transactions on Power Delivery, 25(3), 1629-1636.

[7] Zhao T. (2008). On-line monitoring and fault diagnosis technology research on mechanical state of on-load tap-changer. (Doctoral dissertation, Shandong University).

[8] Zhang H F, Hong-Zhong M A, Chen K, et al. Fault Diagnosis of Power Transformer On-load Tap Changer Based on EMD-HT Analysis of Vibration Signal[J]. High Voltage Apparatus, 2012.

[9] Simas E.F., de Almeida L.A.L., de Lima A.C.. Vibration monitoring of on-load tap changers using a genetic algorithm. In: strumentation and measurement technology conference, 2005. Proceedings of the IEEE, vol. 3, 16–19, May 2005,pp. 2288–93.

[10] Du Y, Wang H, Shi H, et al. Wavelet based soft threshold denoising for vortex flowmeter[C]// IEEE Instrumentation & Measurement Technology Conference. 2009.

[11] Liu T , Zhao T , Zou L , et al. On-Line Monitoring Method of Drive Motor Operational for On-Load Tap Changers[J]. Applied Mechanics and Materials, 2013, 475-476:59-62.

[12] Kay A G . The monitoring and protection of on load tap changers[C]// IEEE Colloquium on Condition Monitoring & Remanent Life Assessment in Power Transformers. IET, 2002.

[13] Wang, C., Lin, L., Ma, H., Jin, J. P., & Zhang, H. (2010). Diagnosis for loose switching contact fault of on-load tap-changer in transformer. China International Conference on Electricity Distribution.

[14] Filho, E. F. S., & Almeida, L. A. L. D. (2008). Self-Organized Classification of On-load Tap Changers Acoustic Signatures. Instrumentation & Measurement Technology Conference

[15] Gao, P., Hong-Zhong, M. A., Zhang, H. F., Chen, K., & Wang, C. N. (2012). Comparison of emd entropy and wavelet entropy in vibration signals of oltc. Proceedings of the Chinese Society of Universities for Electric Power System & Its Automation.

Diagnosis platform development of partial discharge data based on CMMI model

Songyuan Li[1], Chuanghua Liu[2], Xiaoguang Xi[1], Cong Zhao[1], Yangpu Tian[3] and Chengbao Liu[3]

[1] Electric Power Research Institute of Tianjin Electric Power Corporation, Tianjin, 300384, China
[2] Tianjin Electric Power Corporation, Tianjin, 300232, China
[3] Red Phase INC, Xiamen, 361005, China
lisongyuan1987@163.com

Abstract. Capability Maturity Model Integration (CMMI) model is an international standard in software development process and quality management. It integrates the idea of project management body of knowledge and total quality management. Through continuous process improvement for CMMI model, a flexible, regular and progressive capability model framework is established to increase the transparency and controllability of project management and improve the overall effectiveness of software development, which not only ensures software quality, but also solves the bottleneck of traditional project management applied in information technology (IT) industry. This paper presents the analysis and application of CMMI in partial discharge data diagnosis platform, which illustrates its outstanding capability of IT project management. CMMI model make the process manageable and measurable so that the main factors that lead to poor quality and low productivity can be controlled or eliminated.

1. Introduction

According to *condition-based maintenance based on power-on state* from operation and maintenance (OM) department of state grid, information, automation and intelligence of live detection need to be further enhanced[1]. However, the uncontrollable OM process, OM data, test result and personnel allotment limit its development. To realize intelligent OM system, professional OM management, excellent OM benefit and reliable equipment operation, a comprehensive solution was proposed by Tianjin Electric Power Corporation through their years of experiences and achievements in OM. That is the project for diagnosis platform of partial discharge data. To ensure it stable and controllable during the process of project implementation, Capability Maturity Model Integration (CMMI) is used to optimize software improvement and make the development process analysable, predictable and controllable.

CMMI was developed by the software engineering institute of Carnegie Mellon University in the United State with the participation of experts in software process improvement, software development and project management all over the world[2]. It is a assessment standard of capability maturity that has been popularized and implemented widely[3-5]. It is mainly used to guide improvement of software development process and evaluation of software development capability. CMMI framework contains many information guides and behavior references that help enterprises promote their process improvement. For example, at the beginning of a project, the project team should create and maintain a planning process to track and manage the work, and ensure that all members understand and follow the process to carry out project activities. It should be strictly determined how to develop and maintain cost, schedule and product estimates in this plan[6,7]. Implementation effects should be judged by collecting and measuring variances from the plan during the process of project, and corresponding corrective methods have to be taken to ensure the activities be implemented as planned according to the cause of the variances. Besides project planning, tracking and monitoring activities, CMMI also involves project risk management, team building, integrated project management and other related activities.

2. Applications of CMMI in the project

2.1. Project estimation

At the beginning of the project for diagnosis platform of partial discharge data, function point estimation method has been implemented based on the work breakdown structure (WBS) of this development. Function points of this development are figured out with comprehensive consideration of technical, environmental and functional complexity. Calculations of technical, environmental complexity are shown in table 1-2, and statistics for functional complexity of modules are shown in figure 1.

Table 1 Calculations of technical complexity factors.

Technical factor	Instruction	Weight	Complexity
TF1	Distribution level	2	0
TF2	Performance requirement	1	2
TF3	User efficiency requirements	1	2
TF4	Internal processing complexity	1	2
TF5	Reuse level	1	1
TF6	Easy-to-install requirements	0.5	1
TF7	Easy-to-use level	0.5	2
TF8	Portability	2	2
TF9	Easy-to-modify level	1	2
TF10	Concurrency requirements	1	9
TF11	Requirements on special safety features	1	5
TF12	Third-party access	1	0
TF13	Special user training facilities requirements	1	0

Table 2 Calculations of environmental complexity factors.

Environmental factor	Instruction	Weight	Effect
EF1	UML Proficiency	1.5	0
EF2	Application experience	0.5	3
EF3	Object-oriented experience	1	3

EF4	Analyst Capabilities	0.5	3
EF5	Team morale	1	3
EF6	Requirement stability	2	1
EF7	Proportion of part-time workers	-1	0
EF8	Difficulty of programming language	-1	0

UC ID	UC Name	Type	Complexity	UCP
UC001	Supervision and Management of Live Detection	UseCase	Complex	15.0
UC002	Standardized Operation of Live Detection	UseCase	Simple Average Complex	.0
UC003	Business Process Control of Live Detection	UseCase	Average	10.0
UC004	Intelligent Control of Live Detection Implementation	UseCase	Simple	5.0
UC005	Intelligent Diagnosis of Live Detection Data	UseCase	Simple	5.0
UC006	Expert database of live detection data	UseCase	Simple	5.0
UC007	Intelligent State Assessment of Electric Power Equipment	UseCase	Complex	15.0
UC008	Supervision and Management of Abnormal State of Electric Power Equipment	UseCase	Simple	5.0
UC009	Portable management of diagnostic module	UseCase	Average	10.0
UC010	Pilot application of new technology	UseCase	Average	10.0

Figure 1. Statistics for functional complexity of modules.

Accurate workload data are obtained based on accumulated productivity data and estimated results of functional points to provide a reference for cost and schedule estimation. Estimated data of workload and period are shown in figure 2.

1. Project Productivity:

	Productivity	Unit
Average productivity:	7.00	Person-hou/UCP

2. Project Overall Estimation: ◎ 1 Person-months=22 Person-day，1 Person-day=8 Person-hour

Change rate of requirement(RV%)	productivity	UCP	UCP×(1+RV%)	Person-hour	Person-day	Person-months
10%	7.00	463	509	3565	446	20

Workload type	Project R&D	PM	QA	CM	Other	Total
Proportion of R&D		10%	5%	3%	5%	
Workload(Person-day)	446	45	22	13	22	548

3. Estimation of project schedule:

Phase	Requirement Development	High-Level Design	Detailed design	Coding and testing	Integration testing	System Testing	Confirmation and acceptance	Total
Reference upper limit	25%	5%	25%	50%	5%	10%	2%	
Reference Average	20%	3%	20%	40%	5%	5%	2%	
Reference Lower Limit	15%	0%	20%	35%	0%	5%	2%	
Proportion of this project	25%	0%	16%	40%	2%	12%	5%	100%
Workload(Person-day)	137	0	88	219	11	66	27	548
Number of personnel	3	0	4	5	3	4	3	22
Time limit for a project(Day)	46	0	22	44	4	16	9	141

Figure 2. Statistics for functional complexity of modules.

2.2. Project plan

Project schedule, key milestones and other related sub-plans such as communications, project tracking and monitoring, data collection and analysis plan of project and review plan are developed based on estimated workload data. Moreover, this plan need to be reviewed and confirmed by related personnel. Gantt chart of the plan is shown in figure 3.

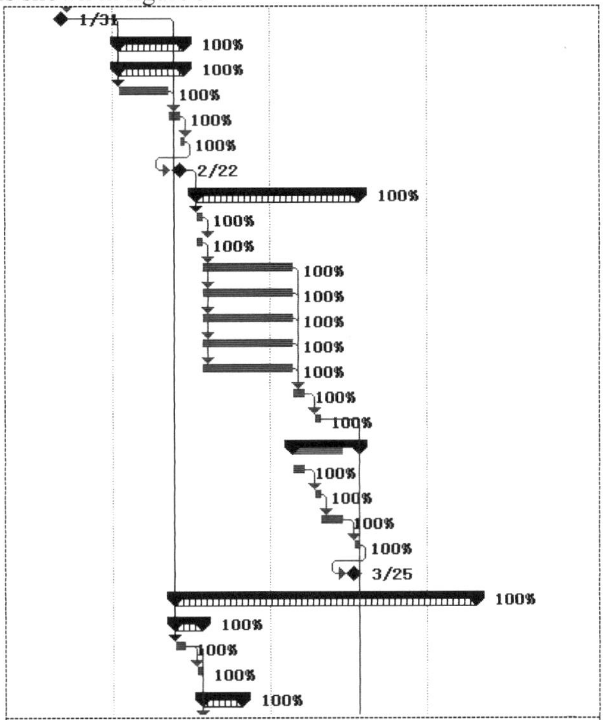

Figure 3. Gantt chart of the plan.

Besides schedule plan, the project plans contain composition and human resource plan (including training plan), project tracking and monitoring plan, measurement and analysis plan, stakeholder participation plan and other plans.

2.3. Tracking and monitoring

The project is tracked and monitored by weekly and milestone ways. Earned value analysis (EVA) method is used to track progress and cost implementation of the project during the milestone period.

In tracking and monitoring activities of the project, manager would compare the progress with the plan so that corrective measures would be taken to avoid problems and risks. Earned value (EV) tracking data in the milestone period are shown in figure 4, and the risk tracking data are shown in figure 5.

1. Analysis of Project Progress

Earned Value Analysis

Current progress of the project	Milestone phase	Accumulate to the present	Target Control Scope		Analysis results	Milestone phase	Accumulate to the present
			lower limit	Upper limit			
Planed Value(PV)	617	1716.0	—	—	Schedule	normal	normal
Earned Value(EV)	620	1727.0	—	—			
Actual Cost(AC)	622	1749.0	—	—			
Schedule Performed Index(SPI)	1.00	1.01	0.80	1.08	Cost	normal	normal
Cost Performed Index(CPI)	1.00	0.99	0.80	1.08			
Schedule Variance rate(SV%)	0%	1%	-10%	10%			
Cost Variance rate(CV%)	0%	-1%	-10%	10%			

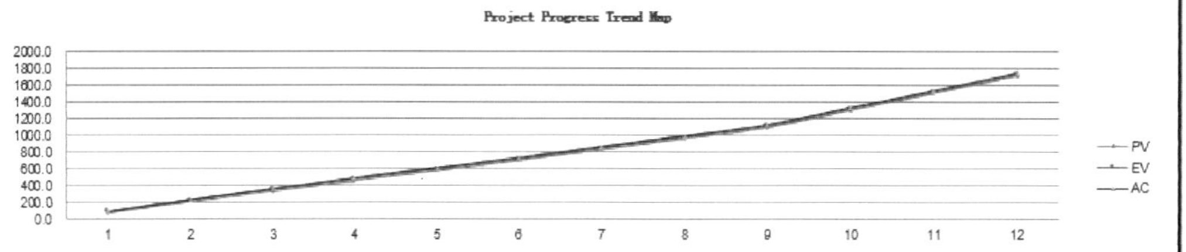

Projects Progress Forecast	Schedule(day)			Effort(Person-hour)			Result analysis
	Plan	predicted value	Time deviation	Plan	predicted value	Effort deviation	
Next Milestone phase	55	55	0	1734	1740	-6	The cost will exceed 6 Person-hour
total	120	119	1	1822	1845	-23	The total cost will exceed 32 Person-hour

Figure 4. EV tracking data of the milestone period.

2. Project risk

Risks and Countermeasures	Risk description	Possibility	Seriousness	Risk coefficient	Risks and Countermeasures
Risks that have occurred	System analysis and design work because of the tight project duration, resulting in simple design, not rigorous analysis and overall consideration, to the late code implementation brings the risk of rework.	4	4	16	The project team strengthens the supervision and evaluation of design, and enables developers to fully understand the importance of pre-design work.
Possible follow-up risks	With the development of the project, the implementation of system functions lags behind schedule	4	4	16	Do a good job of detailed project work decomposition, strengthen the monitoring of work results, to ensure that the system functions are completed on time and in good quality.

Figure 5. Risk tracking data.

At the end of each milestone, process and product audit of quality assurance (QA) is carried out to verify whether the project is implemented following CMMI. When there were variances found by QA, its solving process would be tracked to ensure consistency between process and product[8]. Audit checklist for monitoring process is shown in figure 6.

Project Monitoring and control (PMC)		Project name		Large Data Diagnosis Platform for Partial Discharge	
		Check name		Zhidong Liang	
ID	**Checkpoints**	**YES**	**NO**	**not applicable**	**Notes**
1	Project status tracking				
1.1	Whether project members finish their tasks in accordance with the Project Plan?	√			
1.2	Do project members fill in their personal logs?	√			
1.3	Does the Project Manager Verify the Data in the Personal Log of Project Team Members?	√			
1.4	Does the project manager form Project Weekly after the weekly personal log and before the weekly meeting?	√			
1.5	Does the project manager track weekly non-conformities, defects, and other issues?	√			
1.6	Does the project manager follow up the implementation of the project plan by project members and consult QA Weekly and CM Weekly?	√			
1.7	Does the project manager complete the collection of relevant metrics data and store the collected data in the Project Metrics Database?	√			
1.8	Does the Project Manager Generate the Milestone Report Based on the Analysis and Summary of Data in the Project Metrics Database?	√			

Figure 6. Audit checklist for monitoring process.

2.4. Risk management

Risk management of CMMI has several steps as follows:

- Preparing risk management
- Identifying and analyzing risks
- Reducing or transferring risks

2.4.1. Preparing risk management. The source of risk must be cleared in advance, such as external and internal risks, then the risk coefficient should be determined based on the probability of risk and the severity of the impact. Different management strategies need to be taken according to the risk coefficient[9]. For example, detailed risk management and avoidance plan are required if risk coefficient is high. Proper corrective measures are required if risk coefficient is medium. Tracking with no actions is required if risk coefficient is low.

2.4.2. Identifying and analyzing risks. Possible risk factors of the project should be identified at the start-up stage based on the collected project information, quality and product requirements. Furthermore, the risk, its coefficient and influence scope also need to be evaluated and analyzed.

2.4.3. Reducing or transferring risks. According to the risk management strategy, corresponding risk solutions should be implemented. The impact of risks is tracked weekly to ensure that all risks have been identified and controlled to avoid the impact of risks on the project.

2.5. Peer review

Strict peer review mechanism is adopted in requirement, design and development stage of the project to ensure product quality. For example, firstly, the review experts have received and previewed the software specifications a week before the review. Secondly, deep and intense discussions were conducted among industry experts during the review, and the consensus and problems are recorded. Finally, requirement changes have been decreased due to rigorous and adequate peer review during development process of diagnosis platform.

2.6. Software test

Software testing is a process used to promote and verify the correctness, integrity, security and quality of software products[10]. A detailed software test management plan which covers the requirements of test scope, test strategy, test method, test access conditions and exit conditions, has been formulated in the project planning stage. Test plans and cases must be reviewed before they are implemented.

Besides software test, requirement tracking matrix (RTM) is also used in requirement management activities. The implementation of each requirement can be tracked by RTM, and several specific examples are as follows:

- Whether the corresponding requirement function points are designed, developed and tested
- The impact of requirement change on related products
- Scope of impact on planning, design and other activities.

3. Conclusion

The project for diagnosis platform of partial discharge data is implemented and launched successfully because of the adoption of CMMI module. Corresponding measures can be taken in advance, and the whole process of the project can be clearly guided and supported. This project has accumulated good implementation experience and best practice cases, which can provide reference for the development of similar platforms.

References

[1] Dennis M 2009 *CMMI Essence (3rd edition)* (Beijing: Tsinghua University Press)
[2] Schwalbe K 2011 *IT Project Management* (Beijing: Machinery Industry Press)

[3] Anonym 2017 *A Guide to the Project Management Body of Knowledge (6th edition)* (Washington: Project Management Institute,Inc)
[4] Pritchard C 2004 *The Project Management Communications Toolkit* (London: Artech House Publishers)
[5] Forsberg K. 2011 *Visualizing Project Management* (Beijing: Publishing House of Electronics Industry)
[6] Pınar Ö, Deniz G, Yasemin Y K, Özgün O M and Murat Y 2018 *3rd Int. Conf. on Computer Science and Engineering* (Sarajevo: IEEE) pp 258-263
[7] Wenjin W, Xiaochuan L, Ribo J, Fang L 2017 *3rd IEEE Int. Conf. on Computer and Communications* (Chengdu: IEEE) pp 2508-2511
[8] Muhammad N R 2017 *Int. Conf. on Information and Communication Technologies* (Karachi: IEEE) pp 90-96
[9] Dong-Seok O, Min-Jae L 2016 *7th IEEE Int. Conf. on Software Engineering and Service Science* (Beijing: IEEE) pp 345-348
[10] Sahil K A, Vikas D, Robin S 2014 *Int. Conf. on Advances in Computing, Communications and Informatics* (New Delhi: IEEE) pp 226-230

IOP Conf. Series: Materials Science and Engineering **605** (2019) 012007 doi:10.1088/1757-899X/605/1/012007

The Simulation and Experimental Study of Hydraulic Transmission with Constant-pressure Scheme for Wave Energy Converter Application

S Juwairiyah, M Z Ibrahim*, A Albani, Z M Yusop, M A Jusoh and S Mohad
Eastern Corridor Renewable Energy (ECRE), School of Ocean Engineering, Universiti Malaysia Terengganu, 21030, Kuala Nerus, Terengganu.
*Corresponding Author: zam@umt.edu.my

Abstract. This paper presents the simulation and experimental results of the hydraulic system for wave energy converter application. The different input force applied from the pump unit will imitates the force of wave energy converter (WEC) produced by the ocean waves motion. The simulation conducted using the Hydraulic FluidSIM® 5th version, while the experiment was conducted at Festo-Hydraulic Laboratory. In FluidSIM®, the role of wave floater represented by a set of pump unit that will provide desired pressure to the hydraulic piston. The pump unit consists of a motor, pump, tank and pressure relief valve. The input pressure at the pump unit was adjusted in twelve different levels in the range of 5 Bar to 60 Bar. The different pressure will create a different oil velocity that caused the extension and retraction of the piston. The different pressure also will create a different speed (measured in RPM) of a hydraulic motor. An accumulator and four units of check valves used to drive the moving oil in unidirectional at a smooth flow rate. As a result, the rise of input pressure at the pump unit will increase the speed of a hydraulic motor. For this reason, to achieve the optimal electrical generation by a generator that mounted with hydraulic motor, the higher input pressure is needed. Furthermore, the simulation is validated by the physical experiment to determine the accuracy of generating data. The averaged accuracy of simulation data compared to experimental is 89.92%. After all, the finding from this study will be utilised as a reference in the estimation of the optimal dimension for the actual scale of a hydraulic system for application in the WEC.

1. Introduction

Nowadays, the main problem in the energy sector is the rapid depletion of non-renewable energy resources, such as coal and petroleum. Therefore, the exploration and exploitation of renewable energy resource are the solution to replace the dependent of the energy sector on conventional resource. By utilization of renewable energies such as ocean, wind, solar and other renewable energies will help in slowing the global warming, reducing the carbon emission and sustain the earth's ecosystem. In maritime countries such as Malaysia, that's surrounded by the ocean water, the wave energy is considered has potential for energy generation. This was confirmed by Loon and Koto (2016) that reported that the east coast of Peninsular Malaysia by South China Sea has enormous potential for wave energy exploitation in Malaysia [1].

This paper focuses on simulation and experimental study of hydraulic transmission based WEC system as depicts in Figure 1. In WEC, there are five different PTO configuration for wave energy to electricity conversion including: hydraulic system, air turbine, hydro turbine, direct mechanical drive system linked to rotary generator and direct electrical drive system with linear generator [2]. Although diverse PTO systems have been suggested by different wave energy researchers and developers, hydraulic PTO transmission systems appear to be the choice of the vast majority of developers since there are many WEC device implement this technique [3].The obtained simulation results were then compared with the experiment results. The experimental works of hydraulic power take-off (PTO) system were set up according to the laboratory scale test at Electro hydraulic laboratory facility.

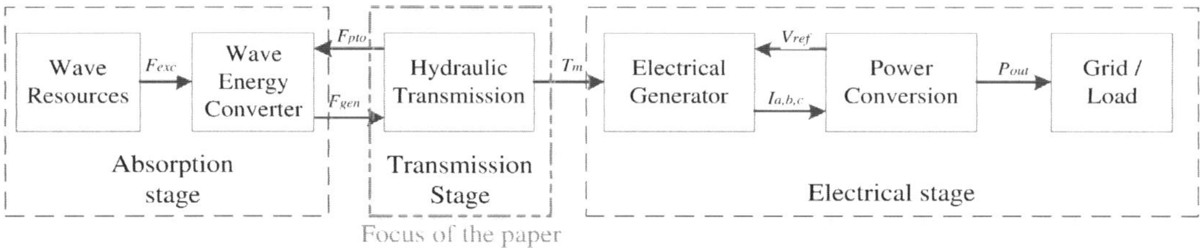

Figure 1. Full conversion change stage in WEC with hydraulic PTO [4].

1.1. WEC device classification

WEC is the device system that harvests the wave energy to generate the electricity. The operating mechanisms of WEC depending on different area conditions; (i) onshore, (ii) shoreline and (iii) offshore conditions [2]. For onshore, the WEC was fixed structure and directly connected to the mainland, for example, the overtopping and oscillating water column. Nevertheless, for the shoreline area, the WEC captured the waves force based on the water depth of the ocean. In offshore area, the mechanism of the device is floater harvester [2]. There are several categories of WECs that differentiated based on their size and direction of elongation; (i) Oscillating Water Column, (ii) Wave Activated Body and (iii) Overtopping [2-3].

Wave activated body (WABs) is the most common and largest application that had been applied in WEC concept [2]. WAB can be classified into two categories; which are floating and submerge. For floating concept with approach of hydraulic PTO, the successive WEC device, Pelamis was mounted and connected with five joint tubes. The WEC was located in semi submerge on the ocean and while the waves come through the Pelamis, the hydraulic power take off used in the system to convert the kinetic energy from the waves force into mechanical energy then convert into electrical energy system [2]. Another floating device with direct mechanical drive as PTO system, UMT Eco Wave Power (UEWP) developed by Universiti Malaysia Terengganu (UMT), Malaysia [5]. The WEC device exploiting the fluctuated movement of floater that connected to the mechanical rack system. In the system, the mechanical rack moving the gear and pulley. The 10 kW generator that connected to the pulley belt will produce the electricity.

For the submerge category of WAB, the PowerBuoy that using a reference plate as a point of reference for the PTO. The device has applied the different solutions for PTO, including the using of oil hydraulics [2].

Figure 2. The P2 Pelamis (left) and PowerBuoy (right) with both use hydraulic PTO as working principle[2]

1.2 Hydraulic PTO in WEC system

In WEC, the Hydraulic PTO systems typically divided into two categories which are variable-pressure and constant-pressure system [4]. For variable-pressure configuration, the hydraulic motor is directly connected with the hydraulic cylinder without intervention of check valve arrangement between the line. The force control is applied directly to the system by means of hydraulic motor will functioned act as pressure regulation and the installation of accumulator will be helpful but are not essential. For constant-pressure system, accumulator implementation is essential to the system in providing smooth output due to highly variable power input. The multiple hydraulic cylinder implementation will affect much this configuration to feasible the force-control. For more details, Costello (2011) reports an analysis of both systems using conventional components [6]. In this paper the analysis will focuses on constant-pressure configuration.

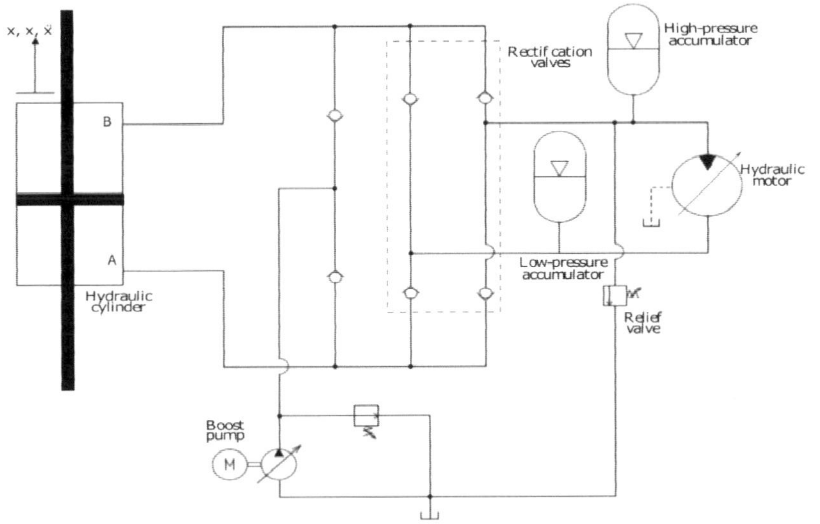

Figure 3. The scheme of a typical constant-pressure hydraulic transmission system [3].

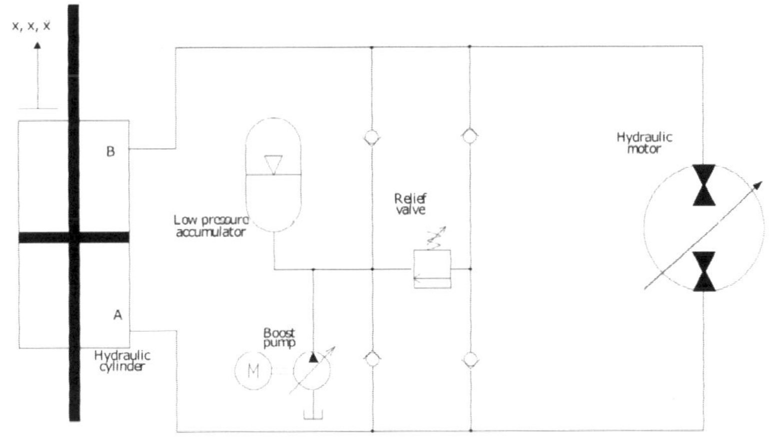

Figure 4. The scheme of a typical variable-pressure hydraulic transmission system [3].

2. Simulation and Experiment

The simulation of hydraulic PTO system was conducted by using the Hydraulic FluidSIM® 5th version. In the simulation set up, a pump unit imitates the designed floater that creates force which acts to the hydraulic double acting cylinder. Furthermore, the main didactic device that had been used for hydraulic

PTO comprises of; (i) coupled of two units of double acting hydraulic cylinders, (ii) a rectifier with the arrangement of four units of check valve, (iii) an accumulator and (iv) hydraulic motor. The components specification while conducting experiment as mentioned in the didactic components depicted in Table 1 and 2.

Table 1. Specification of the hydraulic transmission PTO parameters

Parameter		*Value*
Cylinder	Bore diameter	16 mm
	Rod diameter	10 mm
	Stroke length	200 mm
Accumulator	Volume	0.32 cm^3
	Pre charge pressure	10 Bar
Hydraulic motor	Displacement	8.2 CC/REV
Fluid Oil	Mineral oil	Viscosity 22 cSt (mm^2/s)

Meanwhile, the Figure 5 shows the schematic of the designed WEC system in the simulation interface. The proposed electro-hydraulic system arrangement includes two units of a double acting hydraulic cylinder with both front-end linked together with the coupling nut. It is to allow the second unit of double acting hydraulic cylinder move synchronously when the first double acting hydraulic cylinder excites either extend or retract.

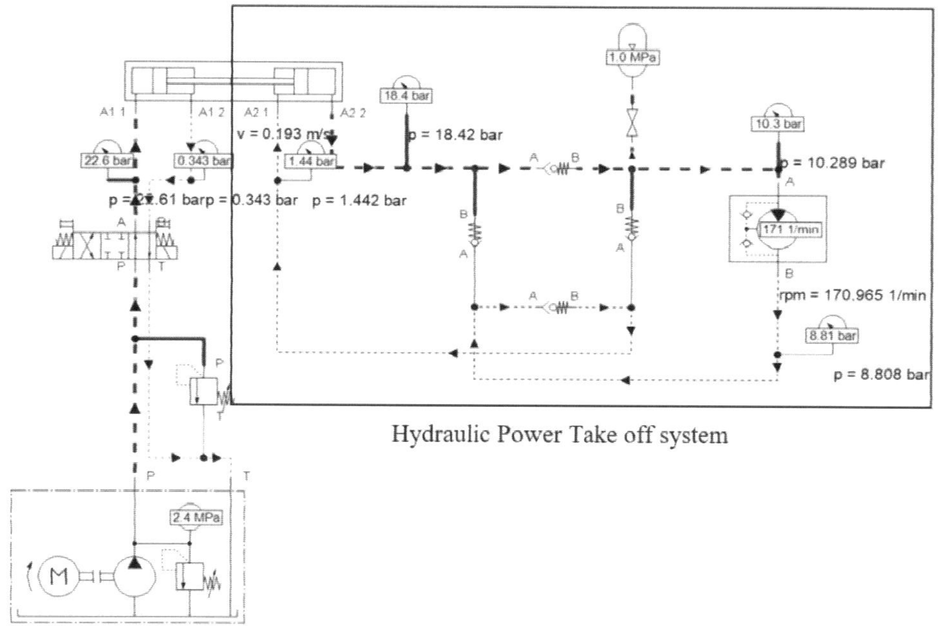

Figure 5. Simulation of the hydraulic power take-off from Hydraulic FluidSIM® 5th version

The experimental works have been set up identically with the simulation circuit design to allow the comparison study. The study on the effects of the hydraulic pressure to the rotational speed of the hydraulic motor was conducted by setting twelve different pressures from 5 Bar to 60 Bar, in the power pump unit of the system. By means of that, the pressure relief valve used in the experiment to allow manually control the input pressure with pressure gauge as an indicator.

The power pump unit as shown in Figure 6 will inject the oil pressure to move the piston inside the first double acting hydraulic cylinder. The power pump and the first double acting hydraulic cylinder will imitate the motions of the waves that move the floater of the system in the real application. The second double acting hydraulic cylinder system acting as the hydraulic transmission PTO of the WEC device. The pressure gauges had been placed at the inlet and the outlet hydraulic cylinder to measure the force input and output of the system. The Digital Tachometer EM2236 was used to measure the rotation of the shaft on the hydraulic motor.

For the second hydraulic cylinder configuration, four units of check valve which assembled in rectified non-return valve and then delivered to the high-pressure accumulator [7]. The purpose of the accumulator is to store the hydraulic energy and deliver the oil flow to the fix displacement type of hydraulic motor and preserve the oil flow rate. The low pressure of oil being drawn after the hydraulic motor passes through the non-return check valve, then will retract hydraulic cylinder continuously. The hydraulic motor functioning to convert the hydraulic energy to kinetic energy and later will pass through the generator and electrical components to produce usable electrical energy.

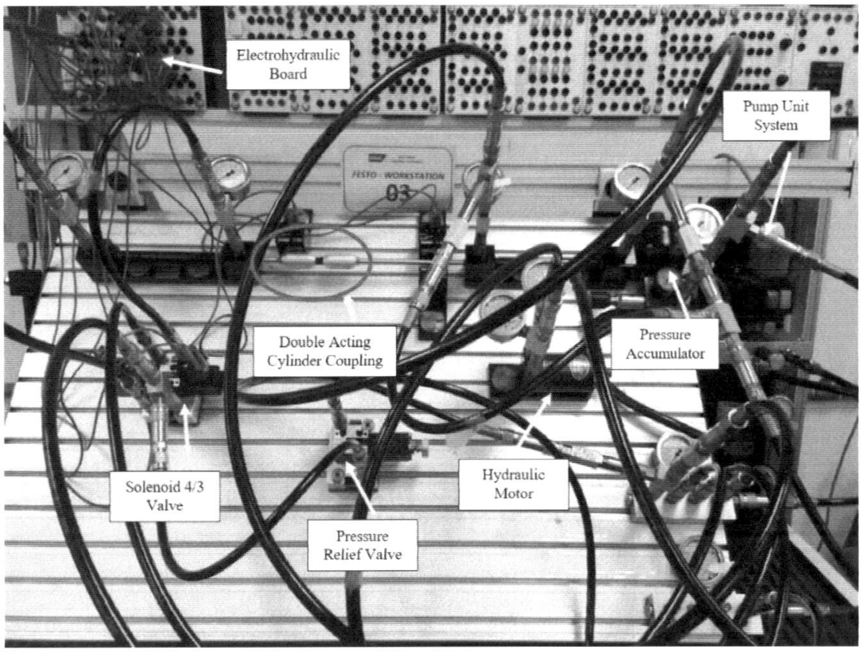

Figure 6. Hydraulic PTO Experiment at FESTO Lab

Table 2. Didactic components in the used in Hydraulic PTO

No	Parts	Symbols	Description
Coupling double acting cylinder		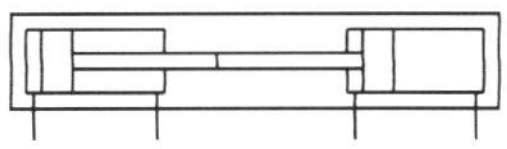	Reacted as Hydraulic PTO connected to the system
Accumulator		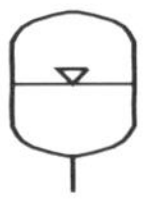	Storage energy for the oil flow
Non return valve	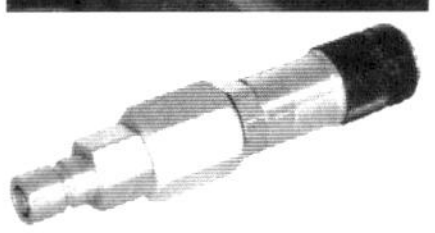		Flow the oil inlet and outlet just in one direction

Pressure Relief valve		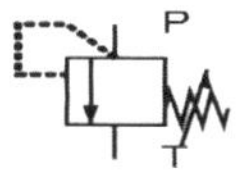	A relief valve or pressure relief valve (PRV) is a type of safety valve used to control or limit the pressure in a system
Hose line	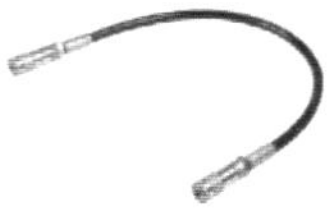	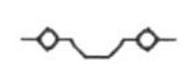	Connection of all components
Hydraulic Motor		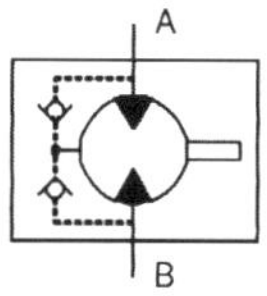	Convert the hydraulic motion into rotational motion
4/3 way solenoid valve		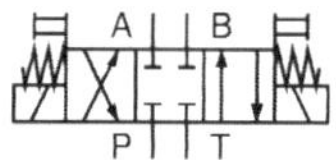	This directional control valve comprises three switching positions for the control of flow rates and is directly actuated via DC solenoid coils.
		Measurement components	
Digital Tachometer EM2236	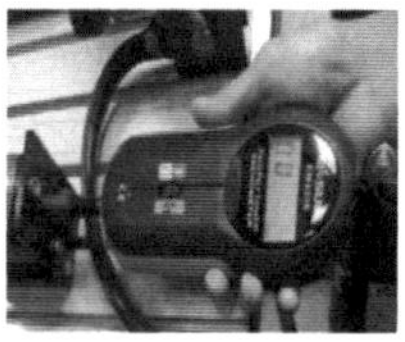		To measure the rotational of the hydraulic motor speed in RPM
Pressure gauge			To measure the pressure at the hydraulic piston.

3. Simulation and Experiment

The physical model of hydraulic PTO system in FluidSIM® simulation is validated by comparing the measurements from the set-up hydraulic PTO test-rig. In this study, the hydraulic test circuit including of variable pump unit and hydraulic actuator represent as the ocean waves motion is utilized to reciprocating the hydraulic cylinder of PTO system. In addition, the hydrodynamics effects on the WEC device are ignored in this study. Therefore, the excitation force, F_{exc} generated from the hydraulic test circuit is directly used as an input of the hydraulic PTO system. Since the F_{exc} is proportional to the pressure setting of pump unit, the F_{exc} is varied by adjusting the pressure of the pump unit from 5 Bar to 60 Bar.

Due to unavailable force sensor in FluidSIM® software and experimental process, the forces are directly extracted from the pressure of the hydraulic cylinder chamber as illustrates in Figure 7 using equation (1) and (2), where, $F_1$and F_2 are rod force and piston force (N), $d_1$and d_1 are rod and piston diameter (m), $P_1$and P_2 are pressure in the cylinder in the rod and piston side, respectively. Table illustrates the extracted F_{exc}, from the hydraulic cylinder of the hydraulic test rig. From the results, there are deviation between the F_{exc} of simulation and experimental may be due to unexpected leakage of hydraulic oil during the experimental validation process. From the table, by adjusting hydraulic pump pressure from 5 Bar to 60 Bar in the simulation and experimental process, the F_{exc} automatically varied from 20 N to 392.1 N and 20.1N to 382 N, respectively.

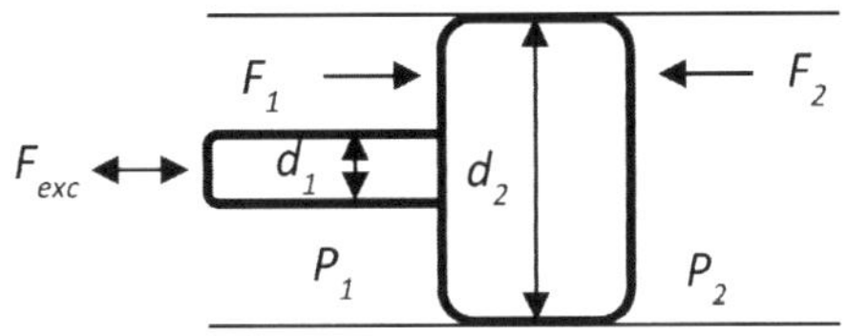

Figure 7. Illustration of force extraction in hydraulic cylinder [8].

$$F_1 = P_1[\pi(d_2{}^2 - d_1{}^2)/4] \quad (1)$$
$$F_2 = P_2(\pi d_2{}^2/4) \quad (2)$$

Table 3. Extracted Excitation Force based on the hydraulic cylinder pressure

Pressure setting of pump unit (Bar)	*Excitation Force, F_{exc} (N)*	
	Simulation	*Experiment*
5	20.0	20.1
10	40.2	40.2
15	140.0	100.5
20	201.0	180.9
25	261.3	221.1
30	291.5	261.3
35	311.6	301.5
40	321.6	321.6
45	331.7	321.6
50	341.8	341.8
55	361.9	361.9
60	392.1	382.0

Figure 8 presents the relationship between the excitation force, F_{exc} and the hydraulic motor speed, HM_ω both from simulation and experimental evaluation results. From the results, the HM_ω is gradually increased proportional to the F_{exc}. However, started from 280 N of F_{exc}, the HM_ω, become constant due to maximum operation limit of hydraulic motor. Based on the results in the Figure 8, the averaged accuracy of simulation model of hydraulic PTO system compared to the experimental is calculated using equation (3), where the $HM_{\omega,sim}$ and $HM_{\omega,exp}$ are the hydraulic speed motor from the simulation and experimental results, respectively. From both analysis, the simulation accuracy is up to 89.92% compared with experiment. There are several factors that influenced the model accuracy during the experimental validation i.e., the pressure losses due to the leakage of hydraulic oil at components connection. In addition, the existence of friction in the piston and hydraulic motor also may be influenced. Apart from that, the inaccurate dynamic parameters setting of the hydraulic components in the simulation model set-up also may affected the performance of the simulation model.

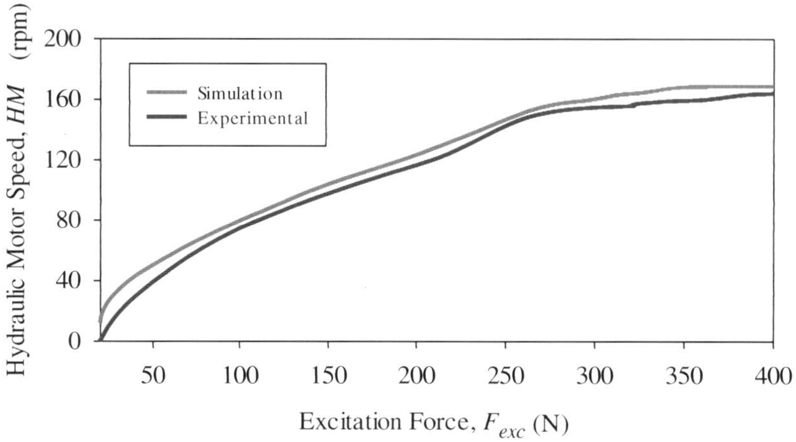

Figure 8. Relationship of excitation force, F_{exc} and the hydraulic motor speed, HM_{ω} from the simulation and experimental evaluation.

$$Model\ accuracy\ (\%) = 1 - \left|\frac{\sum HM_{\omega,sim} - \sum HM_{\omega,exp}}{\sum HM_{\omega,exp}}\right| x100 \quad (3)$$

4.0 Conclusion

In present study, the hydraulic PTO system is developed in high-fidelity FluidSIM® 5th version software. The developed model is then validated through experimental study conducted using hydraulic Festo Lab experimental test rig. The hydraulic test circuit includes of hydraulic pump and hydraulic actuator or cylinder is used to represent the reciprocation motion ocean waves. The pressure of hydraulic pump is adjusted from 5 Bar to 60 Bar to varied the excitation force input of hydraulic PTO system. The simulation and experimental results showed the hydraulic motor speed are recorded at 170 rpm and 160 rpm when the excitation forces are set to 392.1 N and 382 N, respectively. Apart from that, the validation result that obtained from the analysis showed the accuracy of developed simulation model is up to 89.92%.

The results shown significant accuracy between both analysis and lead to crucial outcome for future simulation analysis. Even though, the analysis is done in simulation the system is said capable to emulate the actual behaviour of hydraulic transmission system in real application. The next stage, the simulation analysis will be carried out by setting hydraulic components parameter according to the actual WEC device that currently developed. The hydraulic transmission system needs to properly designed and calculated to ensure the system able to cope the torque and speed rated from electrical generator to produce desired power. For future recommendation, it is also suggested to compare the hydraulic PTO system using Amesim software, which the most used of physical modelling and simulation software for mechatronic systems application.

References

[1] S. C. Loon and J. Koto, "Wave Energy for Electricity Generation in Malaysia -Merang Shore, Terengganu," *Int. J. Environ. Res. Clean Energy*, vol. 304, no. 1, pp. 8–18, 2016.
[2] A. Pecher *et al.*, *Ocean Engineering & Oceanography 7: Handbook of ocean wave energy*. 2017.
[3] M. Penalba, N. P. Sell, A. J. Hillis, and J. V. Ringwood, "Validating a wave-to-wire model for a wave energy converter-part I: The hydraulic transmission system," *Energies*, vol. 10, no. 7, pp. 1–22, 2017.

[4] M. Penalba and J. V. Ringwood, "A review of wave-to-wire models for wave energy converters," *Energies*, vol. 9, no. 7, 2016.
[5] M. D. Lafsah, M. Z. Ibrahim, and A. Albani, "The Development of Wave Energy Conversion Device to Generate Electricity," *Appl. Mech. Mater.*, vol. 773–774, pp. 460–464, 2015.
[6] R. Costello, J. V Ringwood, and J. Weber, "Comparison of Two Alternative Hydraulic PTO Concepts for Wave Energy Conversion," *Ninth Eur. Wave Tidal Energy Conf.*, 2011.
[7] S. Casey, "Modeling, Simulation, and Analysis of Two Hydraulic Power Take-off Systems for Wave Energy Conversion," p. 73, 2013.
[8] "Hydraulic force." [Online]. Available: https://www.engineeringtoolbox.com/hydraulic-force-calculator-d_1369.html. [Accessed: 25-May-2019].

Acknowledgement
The authors would like to thank to the Ministry of Energy, Science, Technology, Environment and Climate Change (MESTECC) and the Electricity Supply Industries Trust Account (AAIBE) for providing a research grant.

Operational Analysis of Institutional Energy System for Developing a Micro-grid

Arvind Sharma[a,b], Mohan Kolhe[a], K.M.S.Y. Konara[a], Kapil Muddineni[b], Ashwini Mudgal[b], Shirish Garud[b], Nils Ulltveit-Moe[a]

[a]Faculty of Engineering and Science, University of Agder. P.O. Box 422. NO-4604 Kristiansand, Norway
[b]The Energy and Resources Institute (TERI), New Delhi – 110 003, India
(Emails: arvind.sharma@uia.no, mohan.l.kolhe@uia.no, konara@uia.no, kapil.muddineni@teri.res.in, ashwini.mudgal@teri.res.in, shirishg@teri.res.in, nils.ulltveit-moe@uia.no)

Abstract— In most of the institutional buildings solar photovoltaic (PV) systems are used for fulfilling the local energy requirement. In this work, a typical institution with solar PV and battery energy storage is selected for analyzing the system's operation to fulfil the institutional energy demand. To operate this energy systemas a micro-grid, distributed generator is selected for fulfilling the institutional load demand in coordination with PV, battery and during the grid outage time. The operation of distributed generator has improved the system performance and coordinated the contribution among local energy resources e.g. PV energy and battery throughput. In this work, relative analysis of the intuitional energy system with and without distributed generator has been analyzed for annual load profile. It has been observed; the distributed generator is not only useful for supplying the energy to the essential load but also during the grid outage conditions. The analysis shows that if, the grid outage continues for longer period then distributed generator can provide the power to the institutional load and operate as a micro-grid. It has been observed, the battery energy throughput as well as effective utilization of local energy resources can be improved by installing distributed generator. The presented analysis is going to be useful for operating the institutional energy systemas a micro-grid with different energy and power management strategies.

Keywords— Battery Energy Throughput, Institutional Hybrid Energy System, Micro-grid, Solar Photovoltaic System

1. Introduction

The penetration of photovoltaic system in the energy network is increasing exponentially. In the distributed network, most of the solar photovoltaic system are going to be building integrated. The photovoltaic system with energy storage can play major role for demand-side management. To overcome the grid outage conditions specially in the weak grid region, PV system with energy storage may require an additional distributed generator. And such system can be operated as a micro-grid with appropriate power and energy management strategies [1]. Therefore, it is very important to select appropriate capacity of distributed generators to meet the load demand especially during the grid outages. In a study [2], techno-economic sizing of an off-grid renewable energy system has been carried out for supplying electricity to the rural community in Sri Lanka however, sizing with grid-connected system is not elaborated with energy pricing. The power conditioning devices of the distributed energy sources can be used to operate the hybrid energy system as a micro-grid (grid connected as well as islanded modes) and can help in regulating the frequency and voltage [3]. The PV system integrated with battery energy storage as well distributed generator can be used as an active generator for operating the entire system as a micro-grid. Such system requires to maintain power quality, reliability and to optimize the energy supply according to the load's characteristics [4]. The electrical energy pricing as well as techno-economic characteristics of the distributed generators have significant impact on the operation of the grid connected energy system [5], and it also affects the battery energy throughput. The grid outage period can significantly impact the operation of PV system and energy storage if appropriate distributed generators are not selected. In the literature, impact of grid outage period on operation of PV system with energy storage is not significantly addressed, and also the analysis of battery energy throughput [6]. The distributed generator can play important role specially during grid outage time, and also improving the battery energy throughput. It has been observed that such analysis has not been sufficiently reported with consideration of real operational results of the institutional hybrid system [7-8].

The main aim of this paper is to analyse operation of the institutional energy system with PV and battery energy storage especially during the grid outage time to fulfil the essential load demand. To improve the performance of the institutional energy system, the role of a distributed generator has been analysed for supplying the power to the load and for improving the battery energy throughput. And, also the importance of energy management strategies for improving the local resources utilization, and how such system can be operated as micro-grid during the grid connected as well as in the islanded mode [10].

In this paper, The Energy and Resources Institute (TERI)'s Retreat Building located at TERI-Gram Gurgaon, India (latitude 28.45 and longitude 77.02) has been considered as an institutional energy system. The TERI-Retreat building energy system has PV capacity of 12.8 kWp, battery (lead-acid) energy storage capacity 48V/ 600Ah and it has been used for providing the electricity to essential loads as well as the main grid to supply the total load demand [11-12]. The block diagram of the existing energy system at TERI-Retreat building has shown in Fig. 1.

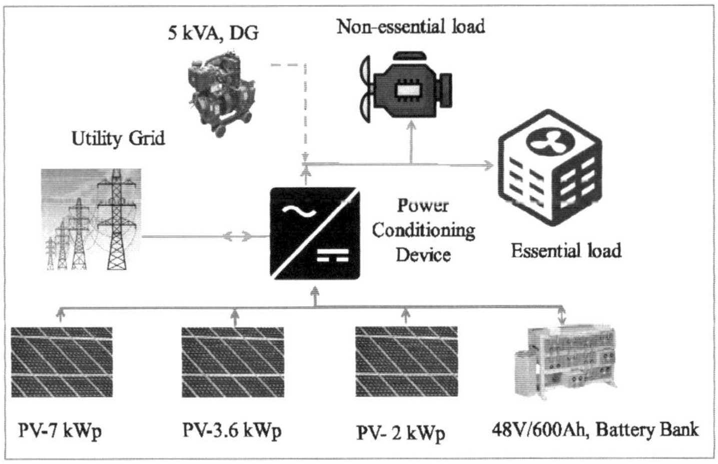

Fig. 1. Block Diagram of existing system at TERI's Retreat Building

In this paper, the operational results of the TERI's Retreat Building energy system has been presented in the section 2. The role of distributed generator during grid outage period with relative performance analysis has been reported in the Section 3. The conclusive results with role of distributed generator for operating the institutional energy system as micro-grid has been presented in the Section 4.

2. Operation Analysis of TERI's Retreat Building Energy System

The schematic of the TERI-Retreat Building energy system is shown in Fig. 1. In this institutional building, electrical load has been divided into the essential and non-essential categories and the total annual load profiles have been presented in Fig. 2. It has been observed that there has been 20% load variation in the daily load profile with reference to the monthly average load profile. The detail analysis of the load profiles has been reported has reported in the ref [11-12]. The PV output profile and load profile has been also reported in the Fig. 2.

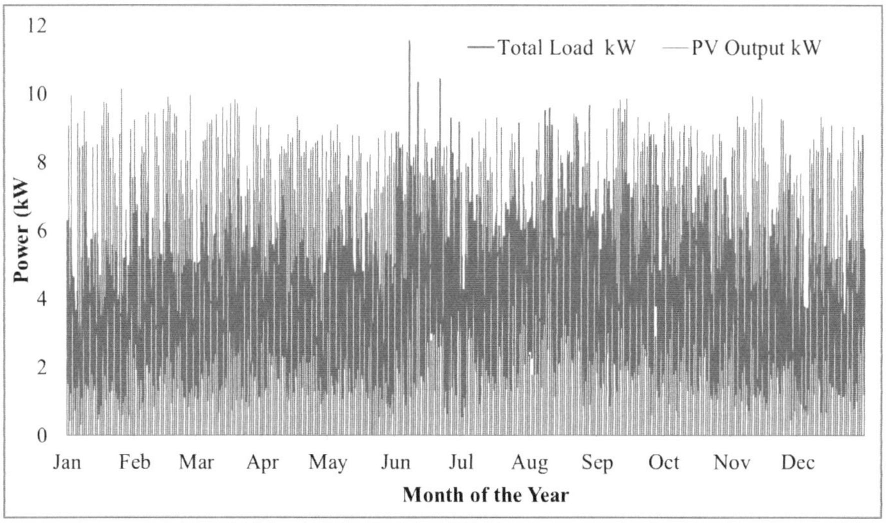

Fig. 2. Annual Profile of the PV Production and the Total Load

It has been observed that the annual energy supply contribution from the grid and the PV for fulfilling the TERI's Retreat building load are 47% and 53% respectively. In the analysis, two hours grid outage has been considered, and especially during this period the role of battery energy storage, for suppling the energy to the essential loads, has been analysed. The monthly average contributions from the PV, grid and the battery energy throughput with monthly average load have been reported in Fig. 3. The analysis has clearly indicated the need of backup power or other distributed energy sources if the grid outage is going to be for longer duration, and it may result in the blackout situation at the institution.

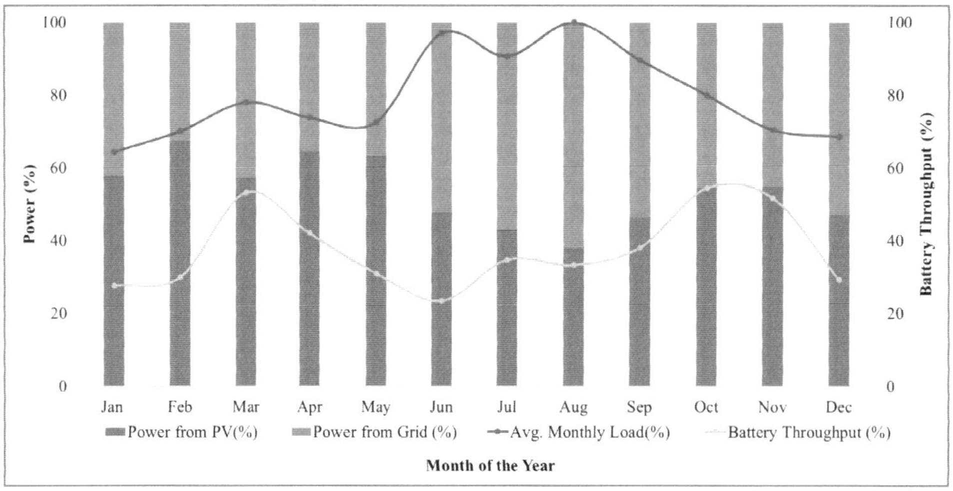

Fig. 3. Monthly contribution from energy sources to the load and battery throughput

The battery operating conditions have been considered with state-of-charge (SoC) from 100% to depth-of-discharge (DoD) of 60%. The lifetime energy throughput of 1 kWh battery has been taken 840 kWh with roundtrip efficiency of 80%.

In the month May (e.g. summer season period), the PV has contributed 63% for fulfilling this monthly load and the remaining 37% has been supplied by the grid with battery throughput of 30%. In the winter season period (e.g. month Dec), the contribution of PV has been 47% and the grid has supplied 53% with battery throughput 29%. The annual average battery throughput has been 37%. The operational results show that if the grid outage is going to increase, then the existing system may not be able to fulfil even the essential load demand. Therefore, it is needed to include a controllable distributed generator not only to avoid the blackout situation, but also to increase the battery throughput for more effective utilization of the PV output; and also, to operate the system as micro-grid during islanded and grid connected mode.

3. Performance Analysis of TERI's Retreat Building Energy System with Distributed Generator

It has been observed that without controllable distributed generator, the TERI's Retreat building energy system has main operational problems during longer grid outage time, and also on the battery energy throughput for storing the excess PV energy (i.e. 3.68%) with appropriate energy management strategies.

In this work, for overcoming the above-mentioned issues, the controllable distributed generator (DG) is introduced without increasing the battery energy storage capacity. The DG can also be operated as controllable source for dispatching the required power as per instantaneous requirement of the institution load with grid constraints as well as during grid outage period [10].

The energy system at TERI's Retreat Building Facility is upgraded by including a DG, whereas the PV and battery capacities are the same as shown in the Fig.1. The schematic of the system with DG has been presented in Fig. 1. Performance of the energy system with DG has been analyzed with the same operating conditions as described in Section 2 and the same typical two months (i.e. May and December) have been selected for analyzing the system operation.

3.1 System Operation Analysis for the Month May

As described in the Section 2, month May has been selected to represent the summer season. During this month May, the contributions from the PV, grid, DG and battery with load have been given in Figure 4 and 5. It has been observed that the monthly average energy contributions to the load from the PV, grid and DG's have been 62.3%, 32.3% and 1.4% respectively; and the DG has operated for 40 hours in the May month. It has been noticed that with DG, the grid contribution has been reduced and the battery throughput has increased to 61%. A typical day 17th of May has been selected to analyse the hourly system operation and observed that especially during the grid outage time, the DG and battery have been contributing effectively and also DG has provided energy to the battery for operating it in the desired conditions. The operational results for 17th May has been given in Fig. 6. On the day 17 of May, the PV, Grid and DG's have contributed 61%, 22% and 17% respectively and the DG has operated three times in a day especially during grid outage periods. The battery energy throughput on this day of 17th May has been 87%.

IOP Conf. Series: Materials Science and Engineering **605** (2019) 012008 doi:10.1088/1757-899X/605/1/012008

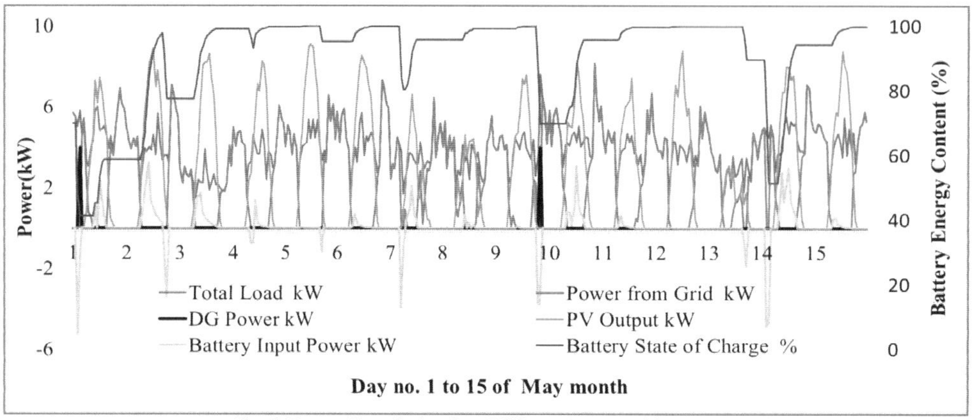

Fig. 4. Energy Contribution from Different Sources of day no. 1-15 of May

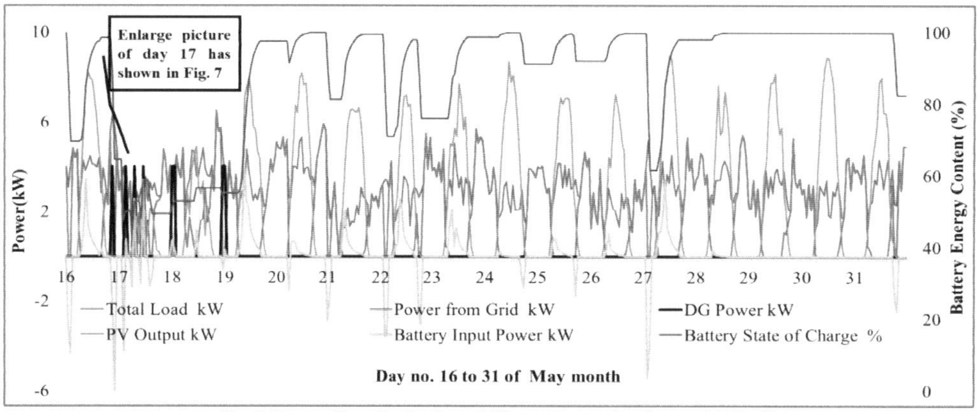

Fig. 5. Energy Contribution from Different Sources for day no. 16-31 of May

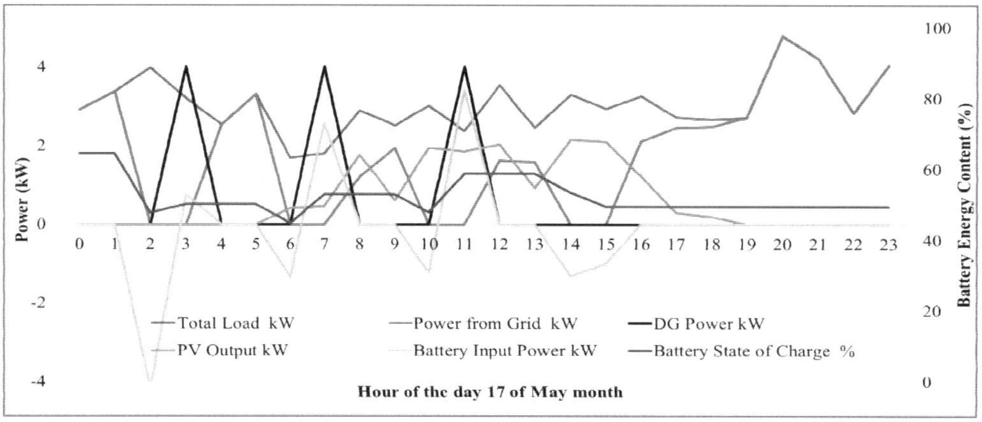

Fig. 6. Energy Contribution from Different Sources on the day no. 17 of May

3.2 System Operation Analysis for the Month December

As described in the Section 2, month December has been selected to represent the winter season. During December month, the contributions from the PV, grid, DG and battery with load have been given in Figure 7 and 8. It has been observed that the monthly average energy contributions to the load from the PV, grid and DG's have been 50.30%, 48.90% and 0.90 % respectively; and the DG has operated for 20 hours in the December month. It has been noticed that with DG, the grid contribution has been reduced and the battery throughput has increased to 73%. as been given in Fig. 7.

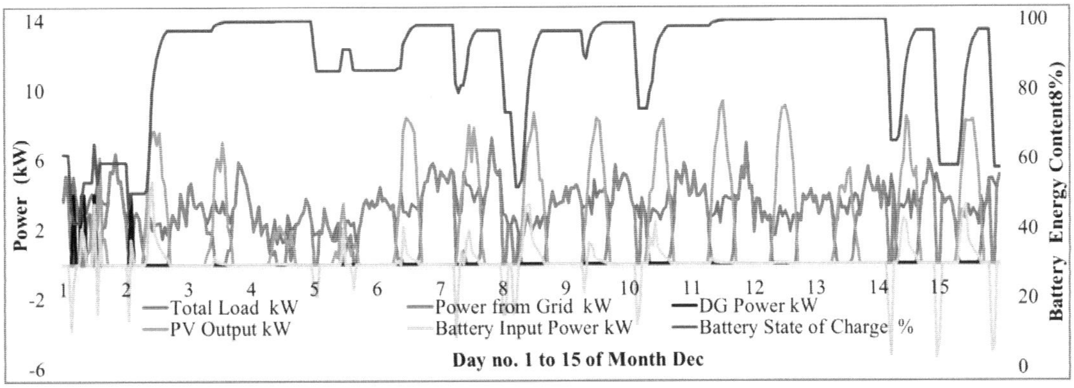

Fig. 7. Energy Contribution from Different Sources for 1-15 days of Month December

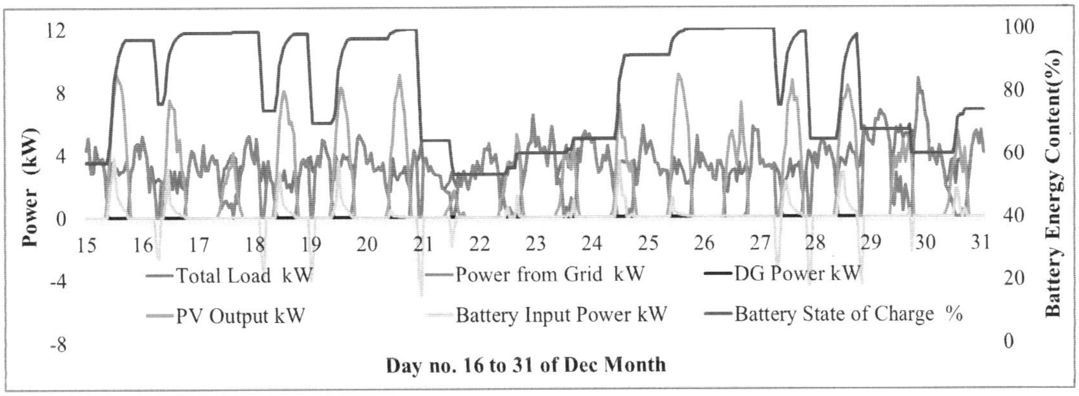

Fig. 8. Energy Contribution from Different Sources for 16-31 days of Month Decemlber

It has been observed from Fig. 7 and Fig. 8, that DG operates only 20 hrs in the month of December and its contribution is estimated 0.9% to the load demand. Except on day 1 and 2, DG doesn't operate on the remaining days of the month as PV along with battery have enough power to provide back up during outage condition. On the day 1, when grid outages occurred and PV doesn't produce enough power and battery's SoC also reached below 50% level, then DG starts operating and meet the instantaneous load demand and as well as charge the battery. DG actively participate and provide back-up power to the entire system's load until grid and PV restored back.

It has been observed from Fig. 7, that battery's SoC has reached to the lowest level of 42% (i.e. 2% above the DoD) on the day 1 and its SoC is recovered by the power contribution from DG and PV. Almost each day in December month, battery's charging and discharging have taken place which indicates, battery is actively participating to meet the load demand during grid outages conditions and therefore its throughput has increased from 29% (i.e. existing system) to 73% (i.e. upgraded system) in the December month. It has been noticed that PV has contributed 12% more as compare to the December month to meet the TERI's retreat building demand which results 16% less energy is purchased from the grid in the May month.

3.3 Annual Operational Analysis of TERI's Retreat Building Energy System with DG

The average annual contribution from PV, Grid, DG and battery are analyzed to meet the TERI's Retreat building load demand. Based on the obtained results, from existing case scenario and upgraded energy system at TERI's Retreat Centre, the annual battery throughput has increased from 37% to 72%, and excess power generation has reduced to 1.67% from 3.68% of the total load and system is able to meet total load demand during outage conditions.

It has been estimated that annual average power contribution from PV, Grid and DG, to meet the load demand are 53%, 45% and 2% respectively and annual average battery throughput is 72%. It has been observed that DG operates in each month of the year and its monthly contribution varies in between 1 to 4%. In comparison to the existing energy system at TERI's Retreat building, battery throughput is more than 60% for each month of the year which indicates a significant increase in the battery's charging and discharging process.

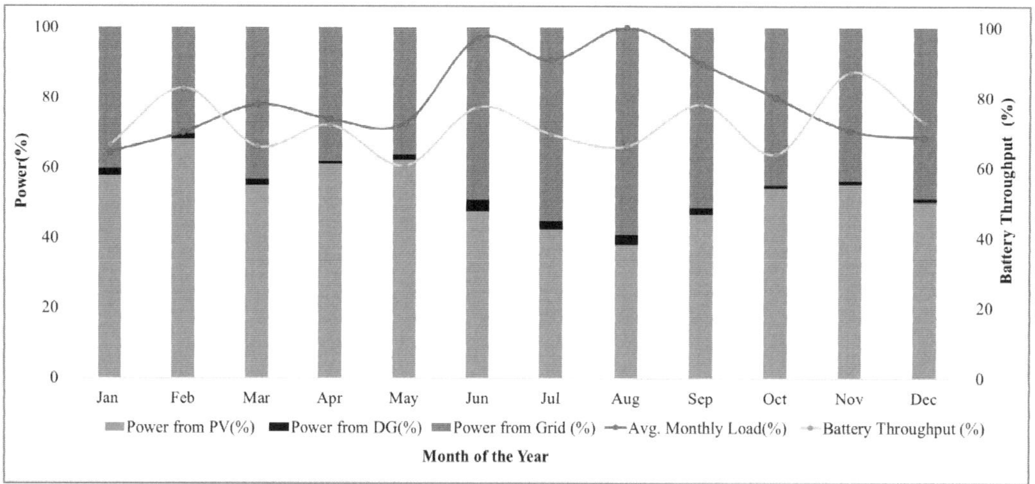

Fig. 9. Monthly contribution from energy sources to the load and battery throughput with DG

It has been observed that even after considering 2 hours random outages throughout the year, 5% more TERI's Retreat Building Load has been fulfilled as compare to existing system. During the year, it has been noted that annual excess power production has reduced to 1.67 % as more PV power is used to charge the battery and therefore increase its throughput. It has been seen from Fig. 3 & 9, in the existing case battery's SoC was always above 60% and average annual throughput was 37% whereas in the upgraded energy system battery's SoC reached to minimum battery level (i.e. 40%) and average annual throughput increased to 72%.

4. Conclusions

In this work, TERI's Retreat Building Energy System as a typical institution is selected for analyzing the system operation to fulfil institutional energy demand. To operate this energy system as a micro-grid, distributed generator is also selected for fulfilling the institutional load demand in coordination with the PV, battery and during the grid outage time. The TERI's Retreat Building Energy System without DG has annual energy supply from the grid and PV for fulfilling the load demand are 47% and 53% respectively with battery energy throughput of 37%. But with the DG, performance of the system with same operating conditions of the grid outage and battery has been improved with PV contribution 53.2%, Grid contribution 45.20%, battery throughput 72%. In this work, DG has main priority to provide power supply to the load during grid outage, but it also ensures the battery charging if, SoC reached below than 50% and other sources are not able to supply the required energy to the load. It concludes that, the upgraded energy system with DG can be performed better compare to the existing energy system and it can also operate as a micro-grid.

Future work is further continuing and focusing on maximizing the local PV production and optimizing the battery energy storage, with appropriate energy management strategy. Energy system performance will be analysed with Time-of-Use Tariff to get maximum economic benefits from battery energy storage. Other distributed energy sources (e.g. biomass gasifier) will be also explored to integrate with the energy system at TERI's Retreat building for providing controllable dispatch power to the micro-grid.

Acknowledgment
This work is supported by the Norwegian Ministry of Foreign Affairs through the Royal Norwegian Embassy, New Delhi under the framework agreement with TERI (India) and University of Agder (Norway) for the project theme 'Sustainability and Clean Energy'.

References
[1] A. N. Azmi, I. N. Dahlberg, M. L. Kolhe, and A. G. Imenes, "Impact of increasing penetration of photovoltaic (PV) systems on distribution feeders," in 2015 International Conference on Smart Grid and Clean Energy Technologies (ICSGCE), pp. 70-74, 2015.
[2] M. L. Kolhe, K. M. I. U. Ranaweera, and A. G. B. S. Gunawardana, "Techno-economic sizing of off-grid hybrid renewable energy system for rural electrification in Sri Lanka," Sustainable Energy Technologies and Assessments, vol. 11, pp. 53-64, 2015.
[3] A. Sharma, M. Kolhe, U. Nils, K. Muddineni, A. Mudgal and S. Garud, "Voltage Rise Issues and Mitigation Techniques Due to High PV Penetration into the Distribution Network," 2018 International Conference on Automation and Computational Engineering (ICACE), Greater Noida, India, pp. 72-78, 2018.
[4] A. N. Azmi and M. L. Kolhe, "Photovoltaic based active generator: Energy control system using stateflow analysis," 2015 IEEE 11th International Conference on Power Electronics and Drive Systems, Sydney, NSW, 2015, pp. 18-22, 2015.
[5] K. M. S. Y. Konara and M. L. Kolhe, "Charging management of grid integrated battery for overcoming the intermittency of RE sources," 2016 IEEE International Conference on Information and Automation for Sustainability (ICIAfS), Galle, pp. 1-6, 2016.
[6] M. Kolhe, K. M. I. Ranaweera, and A. G. B. S. Gunawardana, "Techno-economic analysis of off-grid hybrid renewable energy system for Sri Lanka," in 7th International Conference on Information and Automation for Sustainability, pp. 1-5, 2014.
[7] S. Twaha and M. A. M. Ramli, "A review of optimization approaches for hybrid distributed energy generation systems: Off-grid and grid-connected systems," Sustainable Cities and Society, vol. 41, pp. 320-331, 2018.
[8] M. Lee, D. Soto, and V. Modi, "Cost versus reliability sizing strategy for isolated photovoltaic micro-grids in the developing world," Renewable Energy, vol. 69, pp. 16-24, 2014.
[9] S. Xin, C. Yan, Z. Xingyou and W. Chuanzhi, "A novel multi-microgrids system reliability assessment algorithm using parallel computing," 2017 IEEE Conference on Energy Internet and Energy System Integration (EI2), Beijing, pp. 1 5, 2017.
[10] J. S. L. Senanayaka and M. L. Kolhe, "Integration of distributed energy systems in micro-grid architecture for making a virtual power plant," in 2015 IEEE International Conference on Signal Processing, Informatics, Communication and Energy Systems (SPICES), pp. 1-5, 2015.
[11] A. Sharma, M. Kolhe, U.-M. Nils, A. Mudgal, K. Muddineniand S. Garud, A report on 'Technical Analysis of Real Time Operational Results of Micro-Grid at TERI's Retreat Centre', funded by the Royal Norwegian Embassy, New Delhi under the framework agreement with TERI (India) and UiA (Norway) for the project theme 'Sustainability and Clean Energy'Oct 2018.
[12] A. Sharma, M. Kolhe, K.M.S.Y.Konara, U.-M. Nils, A. Mudgal, K. Muddineniand S. Garud, "Perfomance Evaluation of Institutional Load with Photovoltaic and Battery to Operate it as Micro-Grid" 4th International Conference on Samrt and Sustanaible Technologies (IEEE SpliTech June 2019) (Paper Accepted).

IOP Conf. Series: Materials Science and Engineering **605** (2019) 012009 doi:10.1088/1757-899X/605/1/012009

Load Compensator with an Energy Storage for a Grid Connected PV Based Active Generator

K.M.S.Y. Konara[1,2] , M. L. Kolhe[1] and Arvind Sharma [1]
[1]Faculty of Engineering and Science, University of Agder, PO Box 422, No 4604, Kristiansand, Norway
[2] Faculty of Engineering, University of Ruhuna, Galle, Sri Lanka

E.mails : konara@uia.no, mohan.kolhe@uia.no and arvind.sharma@uia.no

Abstract. At present, in order to build a sustainable economy with less greenhouse emission, the attention has been drawn towards the smart grid concept. In this context, grid connected photovoltaic (PV) systems play a major role. When it comes to grid connected PV systems, the intermittent nature of PV generation and the uncontrolled power injection may cause excess variations in the grid voltage, frequency and the power quality. Therefore, the concept of active generator with innovative energy management strategies is a good alternative with which the PV generated power can be injected to the grid in a controlled manner while achieving the output power quality, instantaneous power, balance, frequency control and voltage control. This work proposes an innovative architecture for a grid connected PV based active generator with multi sources; a PV array, battery and the grid supply which are connected to a common AC bus as well as it analyses the innovative load flow control strategy associated with the proposed architecture. To make sure an uninterruptable power supply to the load, PV arrays inject the generated power in maximum power point tracking mode and the battery compensates the power requirement in coordination with the grid while maintaining the frequency stability within the system. Therefore, this paper analyses the performance of the energy controller and the load management capability of the proposed system while achieving the frequency control.

1. INTRODUCTION

Though photovoltaic (PV) energy stands as one of the prominent sources of renewable energy in present, some problems can arise due to its stochastic nature when increasing the more penetration of PV generated power in uncontrolled manner. The intermittent nature of PV generation and the uncontrolled power injection with passive power conditioning devices can result power quality issues and may cause excess variations in the grid frequency and the grid voltage [1, 2]. Especially, on the day time of a clear sunny day, PV arrays may produce excess energy that can be stored in a local energy storage or injected to the utility grid in addition to supplying locally connected loads. In order to overcome from the aforementioned issues related with the uncontrolled PV integration, PV based active generator (AG) with embedded energy storage can be employed as a solution. But, this type of distribution systems with active generators necessitate innovative energy management and control strategies for increasing the more penetration of PV generated power without getting any power quality issues and frequency fluctuations [1, 3].

Ref [4] analyzes the power management capability of a grid integrated battery for a distributed energy system in order to provide the prescribed power to the grid. They have mainly focused on handling the power quality and correcting the power factor by using an active filter, when it is connected to a distributed energy system. The authors have discussed about eliminating the power fluctuations, but the frequency control and the voltage control are not addressed in this work. The demand management of a PV system with various storage technologies is presented in [5]. In the proposed architecture, multi sources are connected to a common DC bus and the voltage source inverter (VSI) connected to the DC bus has been controlled using current mode control. They have significantly focused on the demand management using different power shares of the multi sources but failed to address the frequency control of the utility grid. Ref [3] has focused on the control and power management of a grid-connected PV system with an energy storage but it has not given a significant attention to manage the frequency variations caused by the integration of intermittent PV generated power. Although considerable research efforts have been devoted to energy management of a grid connected PV based AG, less attention has been paid to control the frequency fluctuations result from the integration of the AG with the utility grid.

Most of the grid connected PV systems are based on the architecture shown in Figure 1. The VSI connected to the common DC bus is controlled using current mode control. The required current reference values of each power source in the AG are calculated using different mathematical models developed for each component of the AG. It can be observed that, the independent control of individual power sources has become inconvenient with this architecture.

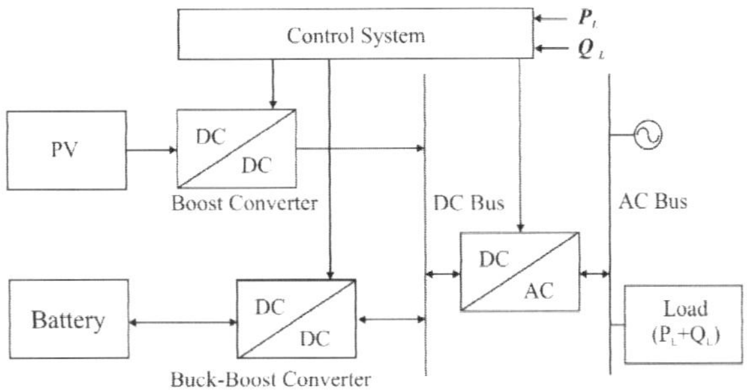

Figure 1. System architecture of the PV based AG

Therefore, this work proposes an architecture for a PV based AG with a common AC bus to which all the power sources are connected along with their own energy control system while maintaining the frequency control of the grid. To maintain reliable and healthy operation of the system, the availability and the accessibility of the battery should be taken into account when proposing a proper load management algorithm for a PV based AG system. Since the proposed system consists of two power sources, PV array and battery, a hierarchical approach with PID control and state-flow control are used for power dispatching from different sources. In this work, Matlab/Simulink® is used to model the power sources and the power conditioning devices and the logic controller for energy management is achieved with Stateflow® model.

2. The proposed architecture of the PV based AG and its components

Generally, a PV based AG consists of multi power sources which dispatch power to meet the load demand, power conditioning devices, locally connected loads and the energy control system. Basically, four different sub systems can be identified for designing a PV-AG; 1. Power generation subsystem 2. Power conditioning devices subsystem 3. Energy storage subsystem 4. Energy management subsystem. Depending on the power dispatching strategies and the constraints enforced by the utility grid, suitable components including power electronic converters and energy storage technologies are selected to the mentioned subsystems of the PV-AG.

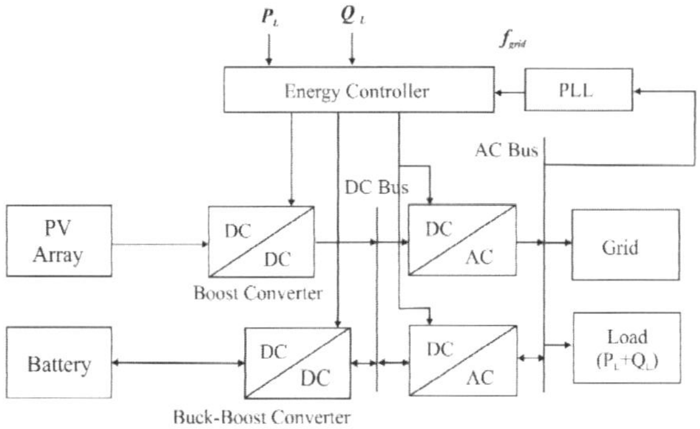

Figure 2. Proposed system architecture of the PV-AG for load management and frequency control

The proposed system architecture of the PV-AG for load management and frequency control is shown in Figure 2. The mentioned subsystems of the PV-AG are connected to a common AC bus to which the utility grid is connected. In order to achieve proper operation as an AG, power flow among the PV array, energy storage sub system, locally connected load and the grid should be properly coordinated.

2.1. Power generation subsystem

In this work, a 215 kW (peak) solar PV array is selected. The array consists of 54 strings in parallel and each string is having 20 panels in series. The specifications of Monocrystalline Solar Panel Merkasol 200W PV panels manufactured by Merkasol® [6] are employed to model the PV array output at the Maximum power point (MPP). The relevant technical data of the selected solar PV module is tabulated in Table 1.

Table 1. PV module data [6]

Electrical data	Parameters
Nominal output P_{mpp}	200 W
Nominal Voltage V_{mpp}	36.9 V
Nominal current I_{mpp}	4.55 A
Open circuit voltage V_{oc}	45.2 V
Short circuit current I_{sc}	5.72 A
Conversion efficiency	15.67%

To analyze the performance of the proposed demand management strategies, it is necessary to know the PV array output. The output of the PV array depends on the incident solar radiation, module temperature and the parameters of the cell. In this work, the variation of global tilted irradiation (GTI) data is considered as illustrated in Figure 3.

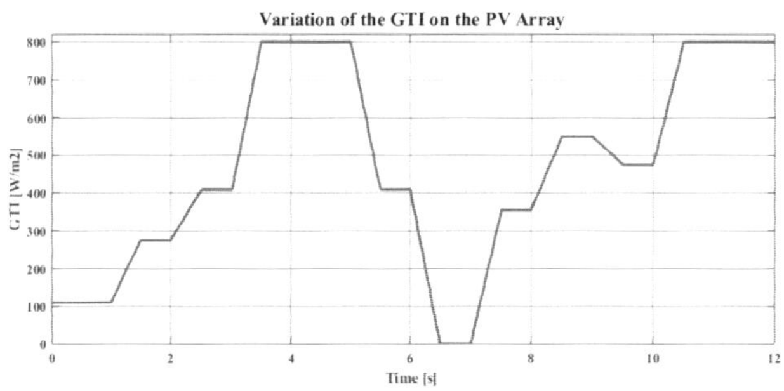

Figure 3. GTI variation on the PV array

2.2. Energy storage subsystem

In the proposed PV based AG system, lead-acid battery is used as the long-term energy storage. The battery capacity with reference to the load and the PV capacity has been selected based on the study presented in [3, 7, 8]. When the PV generated power is not sufficient to meet the load demand, the battery discharges its power at lower rate. The battery should maintain a high state of charge (SOC) by charging when excess energy is available within the system. In this work, the battery model available in MATLAB/Simulink® is used to implement the battery subsystem. The battery model is configured as a lead-acid battery with the parameters given in Table 2.

Table 2. Parameters of the battery

Parameters	Data
Nominal Voltage	400 V
Rated Capacity	200 Ah
Initial State Of Charge	80 %
Fully Charged Voltage	435.6 V
Nominal Discharge Current	13.9 A
Internal Resistance	0.02 Ω
Capacity(Ah)@Nominal Voltage	62.1Ah

2.3. Power conditioning devices subsystem

The system consists of one DC-DC boost converter, one bi-directional DC-DC buck-boost converter and two 3φ VSIs as power conditioning devices. In the PV-AG system, the DC-DC boost converter connected with the PV array works in MPPT mode by controlling its output voltage [9]. The bi-directional DC-DC buck-boost converter connected with the lead-acid battery facilitates both charging and discharging modes as it is capable of handling bi-directional regulated power flow. In the VSIs, the DC sides are connected to the two DC-DC converters and the AC sides are connected to the common AC bus to consume and inject the frequency regulated AC power to the common AC bus.

2.4. Energy management subsystem

The energy management system is designed in hierarchical fashion as an event driven system with PID control and state-flow control. Hierarchical approach of controlling includes several hierarchical stages and each stage is responsible for doing a control task based on its hierarchical position [10]. In this design, this system includes four stages each having their own control task depend on the hierarchical position as illustrated in Figure 4.

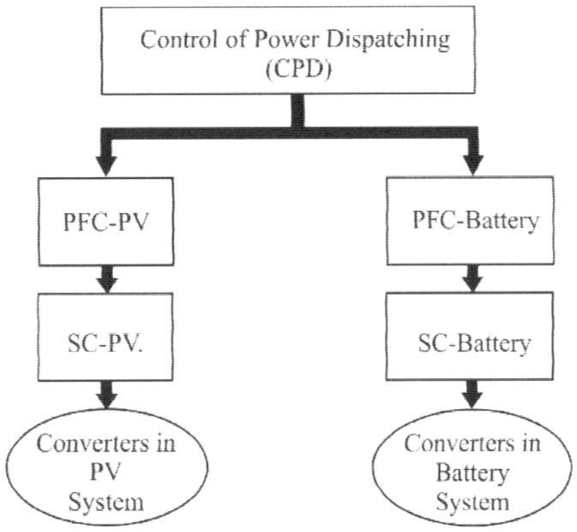

Figure 4. Energy management subsystem overview

3. Energy management and frequency control of the PV-AG

As described in 2.4, the whole control task is subdivided into three hierarchical stages as illustrated in Figure 4. Control of power dispatching (CPD) layer determines the availability and the accessibility of each source and the required power set points to achieve proper power dispatching among the available sources. Power shares are determined at the power flow control (PFC) level to control the system frequency. Based on the reference power signals calculated at the PFC, switching control (SC) level generates the required PWM signals to each converter to deliver the required power to the load.

3.1. Control of power dispatching (CPD)

At this stage, the availability and the accessibility of the considered power source for power dispatching are determined. The control layer is designed as an event driven system with state-flow control. In order to get the decision, SOC of the battery and the generation capacity of the PV array are considered. The state transition diagrams of CPD for PV array and battery are shown in Figure 5 (a) and (b) respectively [10].

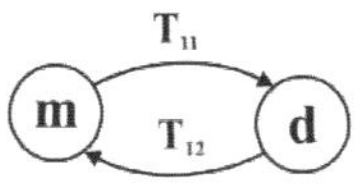

m - MPPT mode
d - Disconnected mode

Figure 5(a). State transition diagram of CPD for the PV array

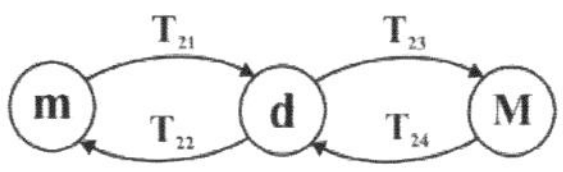

m - Min mode
M - Max mode
d - Dispatchable mode

Figure 5(b). State transition diagram of CPD for the battery

State transition logics of CPD for PV array and battery are tabulated in Table 3. (a) and (b) respectively. The flow chart of the state transition logic of the battery is illustrated in Figure 5 [10].

Table 3(a). State transition logic of CPD for the PV array

State Transition	Logic
T_{11}	$[P_{pv}(t)>P_{pv,min}]$
T_{12}	$[P_{pv}(t)<P_{pv,min}]$

Table 3(b). State transition logic of CPD for the battery

State Transition	Logic
T_{21}	$[SOC(t)>SOC_{min}]$
T_{22}	$[SOC(t)<SOC_{min}]$
T_{23}	$[SOC(t)>SOC_{max}]$
T_{24}	$[SOC(t)<SOC_{max}]$

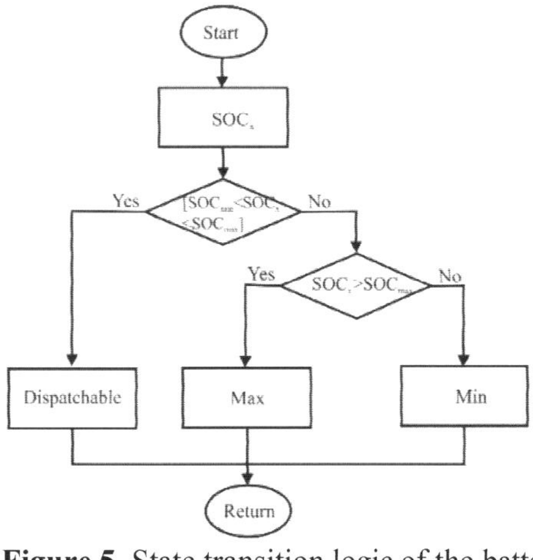

Figure 5. State transition logic of the battery

$P_{pv}(t)$ - current PV generation of the array
$P_{pv,min}$ - minimum PV generation
$SOC(t)$ - current state of charge of the battery
SOC_{max} - maximum allowable state of charge of the battery for dispatching
SOC_{min} - minimum allowable state of charge of the battery for dispatching

3.2. Power flow control (PFC) and PWM switching control (SC)

In the dispatchable mode, the PV array inject the maximum available power in MPPT mode with the help of the closed loop control developed at the PFC layer for PV and then the load demand is compensated by the battery and the grid if these two power sources are determined as accessible or available at the CPD stage. To facilitate the battery in both charging and discharging modes, three P_{set} values are obtained from the CPD-battery stage depending on its current state; *d, m* and *M.* Based on this, the PFC of the battery controls the required power share of the battery while controlling the frequency. According to the required reference current components (i_{d_ref} and i_{q_ref}) of both the power sources calculated at the PFC layer, required PWM signals for VSIs are generated.

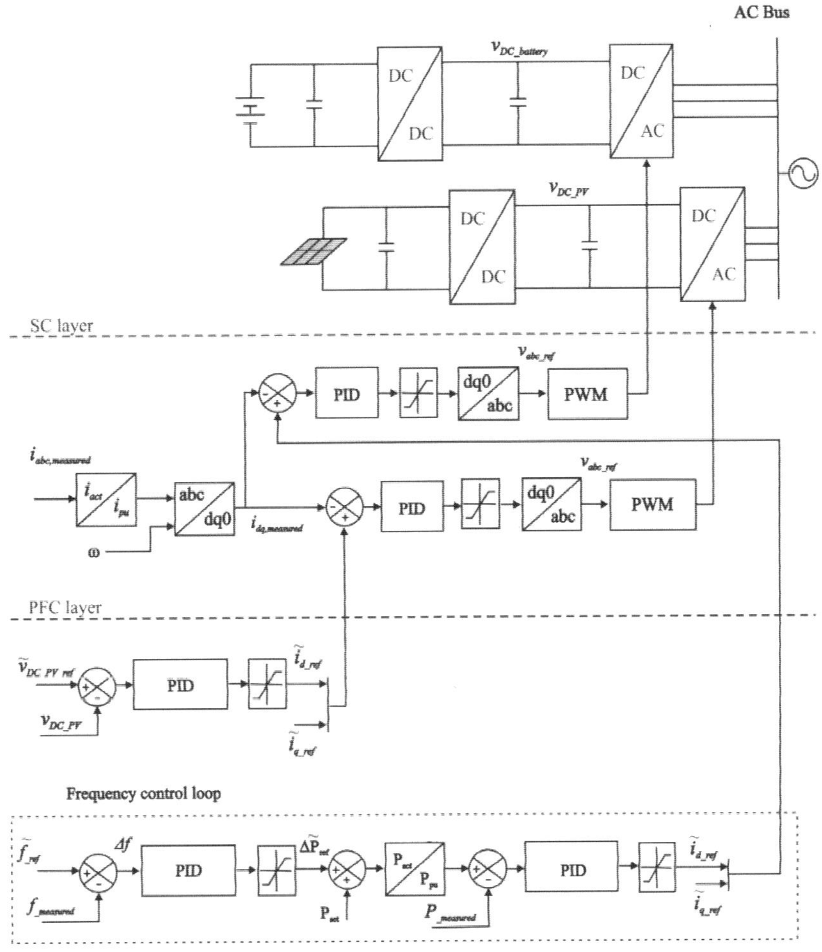

Figure 6. PFC and SC of the PV-AG

4. Results

This section provides the results which analyze the performance of the proposed control strategies of the PV-AG. Energy management capability of the proposed PV-AG architecture is evaluated for the load demand variation illustrated in Figure 7 and the GTI variation shown in Figure 3.

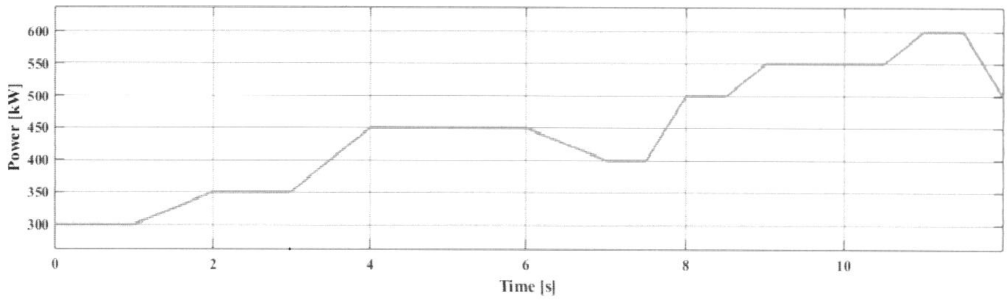

Figure 7. The variation of the load demand

4.1. Sharing the load demand among the multi-sources in the PV-AG

The share of the load demand among each power source is illustrated in Figure 8.

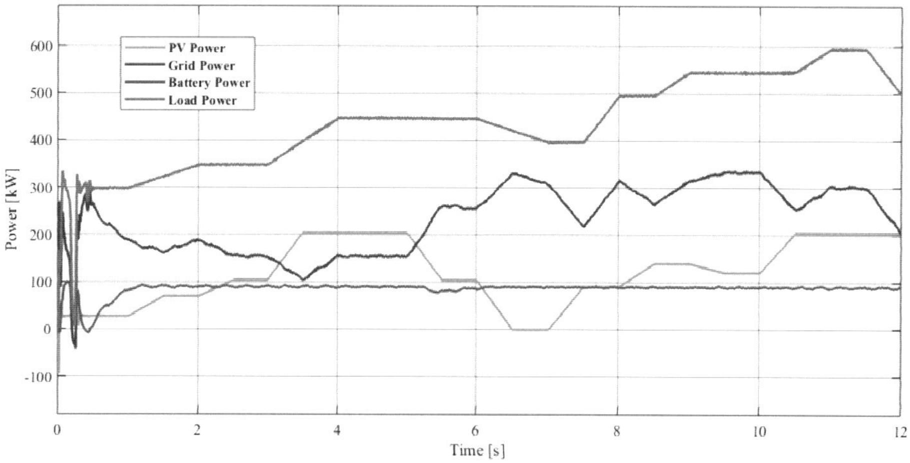

Figure 8. Load demand sharing among the multi-sources of the PV-AG

The PV based AG system starts supplying the load at 0 s. The system undergoes very short period of transients and becomes stable approximately within 0.35 s. It can be seen that the PV array injects the maximum available power to meet the load demand as it follows the GTI variation. The battery and the grid have shared the remaining power of the load demand. The battery has taken more or less constant power share through out the operation as it works as a long-term energy compensator while maintaining the frequency stability within the system. The grid has taken care of all fast dynamics of the load demand.

The variation of the frequency of the AC bus is shown in Figure 9. In this simulation, the grid frequency was configured as 50 Hz. As long as the PV-AG supplies the load demand, it keeps the system frequency within 49.8 Hz to 50.2 Hz with the proposed control strategies.

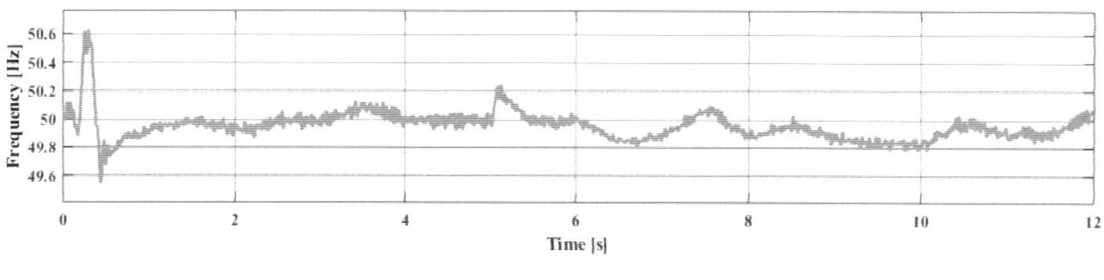

Figure 9. Frequency variation of the grid

5. Conclusion

When increasing the penetration of PV energy, power system stability issues will be arising due to the intermittent nature of PV generation. The PV based active generator (PV-AG) is an effective solution for increasing the penetration of PV produced energy without affecting the stability of the power grid. In this work, an AC coupled architecture for a PV-AG in which all the power sources (PV array, lead-acid battery) are connected to a common AC bus through the separate DC-DC and DC-AC conversion stages is developed and implemented as a simulation model. With the developed innovative control strategies for the proposed PV-AG architecture, the load demand of the AG system can be properly managed and controlled while maintaining the frequency stability within the system. With this proposed architecture, short-term load variations are handled by the utility grid. Therefore, as a future work, the system could be improved to handle the short-term fluctuations by the system itself with the lesser involvement from the grid.

6. Acknowledgement

The University of Agder (UiA) gratefully acknowledge the partial support extended by the Norwegian Ministry of Foreign Affairs under the framework agreement with The Energy and Resources Institute (TERI) (India) and UiA for the project theme 'Sustainability and Clean Energy'.

References

[1] M. H. d. F. Takami, S. A. O. d. Silva, and L. P. Sampaio, "Dynamic performance comparison involving grid-connected PV systems operating with active power-line conditioning and subjected to sudden solar irradiation changes," *IET Renewable Power Generation,* vol. 13, no. 4, pp. 587-597, 2019.

[2] M. Kolhe, "Techno-Economic Optimum Sizing of a Stand-Alone Solar Photovoltaic System," *IEEE Transactions on Energy Conversion,* vol. 24, no. 2, pp. 511-519, 2009.

[3] A. N. Azmi and M. L. Kolhe, "Photovoltaic based active generator: Energy control system using stateflow analysis," in *2015 IEEE 11th International Conference on Power Electronics and Drive Systems*, 2015, pp. 18-22.

[4] V. Virulkar, M. Aware, and M. Kolhe, "Integrated battery controller for distributed energy system," *Energy,* vol. 36, no. 5, pp. 2392-2398, 2011/05/01/ 2011.

[5] D. Lu, T. Zhou, H. Fakham, and B. Francois, "Design of a power management system for an active PV station including various storage technologies," in *2008 13th International Power Electronics and Motion Control Conference*, 2008, pp. 2142-2149.

[6] (5/4/2019). *Monocrystalline Solar Panel Merkasol 200W 24v High Efficiency* [Online]. Available: https://www.merkasol.com/Monocrystalline-Solar-Panel-Merkasol-200W-24v-High-Efficiency.

[7] M. A. M. Yassin, M. Kolhe, A. Sharma, and S. Garud, "Battery Capacity Estimation for Building Integrated Photovoltaic System: Design Study for Different Geographical Location(s)," *Energy Procedia,* vol. 142, pp. 3433-3439, 2017/12/01/ 2017.

[8] M. A. M. Yassin, M. L. Kolhe, and A. N. Azmi, "Battery capacity estimation for building-integrated photovoltaic system: Design study of a Southern Norway ZEB house," in *2017 IEEE PES Innovative Smart Grid Technologies Conference Europe (ISGT-Europe)*, 2017, pp. 1-6.

[9] K. M. S. Y. Konara, M. L. Kolhe, and A. Nishimura, "Grid integration of PEM fuel cell with multiphase switching for maximum power operation," in *2016 IEEE International Conference on Power System Technology (POWERCON)*, 2016, pp. 1-5.

[10] K. M. S. Y. Konara and M. L. Kolhe, "Charging management of grid integrated battery for overcoming the intermittency of RE sources," in *2016 IEEE International Conference on Information and Automation for Sustainability (ICIAfS)*, 2016, pp. 1-6.

IOP Conf. Series: Materials Science and Engineering **605** (2019) 012010 doi:10.1088/1757-899X/605/1/012010

Analytical Modelling of Power Swing and Validation Using Real Time Digital Simulator

A Zeno[1,2], J R Orillaza[1,3], M L Kolhe[4,5]

[1] University of the Philippines Diliman, Quezon City, Manila, Philippines 1101
[2] aldrichzeno@rocketmail.com, +63-9277607642
[3] jordan.orillaza@eee.upd.edu.ph
[4] Faculty of Engineering and Science, PO Box 422, No 4604, Kristiansand, Norway
[5] mohan.l.kolhe@uia.no

Abstract. Grid disturbance is one of the largest areas of concern in the protection of power systems after high penetration of distributive generation. They are responsible for causing mass blackouts. This paper analytically derives the equations for resistances and reactances as a function of time. The derived expressions were validated using real time digital simulator (RTDS) for variation in power factor of system and frequency of the power swing. Results signify that the derived equations were able to provide the locus of instantaneous values of the real time power swing occurring in the system for changes in power factor and swinging frequency. Two dimensional R-X plane and three dimensional R-X-t space plots were also used to study the static and dynamic variations of impedances in time domain. The expressions r(t) and x(t) will be useful in determining the instantaneous short circuit ratios and also for tuning the impedance relays.

1. Introduction

With the higher penetration of renewable resources, the large inertia of a synchronous machine is constantly getting replaced by smaller or zero inertia renewable generators like solar, tidal or wind. Decentralisation of large synchronous generators with smaller units of distributed induction machines or solar panels has challenged the grids synchronicity and stability. In view of power system dynamics, stability studies play a vital role for the renewable energy integration. Power swing occurs when a disturbance occurring in one part of a system due to absence or limited inertia leads to propagation of disturbance along the whole grid. These swings cause mis-operation of relays especially the distance relays in transmission lines causing regional or country wide blackout [1]. Power swing is an un-attenuated periodic disturbance caused due to oscillations in voltages and currents thus forcing the impedances to swing around the normal operating point which distorts the locus of tripping and restraining region of relays. Thus it is vital to understand power swing in order to maintain grid stability and for integrating large scale renewable generation with conventional centralised generators [2].

The parameters which are observed by measuring meters at any point in time are voltages and currents. Variation of voltages and currents lead to oscillation of active and reactive power which leads to rise of stability studies during power swing. Eigen value analysis [3] helped in finding the swinging frequencies, yet required numerous computations. Transformation of domain from time to frequency [4] provided the magnitude and frequency of swinging frequencies, yet the continuous online transformations and calculations for small time steps seemed to be memory demanding and laborious. Thus, it is essential to have a generalised analytical model which could describe the power swing for any frequencies and power factor which eliminates the technical difficulties and solely depend on time varying few sets of non iterative equations [5].

Here, voltages and currents in time domain will be used for analytically deriving the resistances and reactances as a function of time – viz. r(t) and x(t) during power swing. Evaluation of instantaneous resistances and reactances may have the following functions.

1.1 This will shed light on the amount of distributive generations which could be penetrated into the grid based on voltages, currents and power factor before the system looses stability.

1.2 System strength or stiffness is the ability of the power system to maintain voltage stability, frequency stability and quality at the Renewable Energy Source interconnection points. The stiffness of any grid is measured using short circuit ratio which is a function of resistance and reactance. Instantaneous short circuit ratio x(t)/r(t) will help in accessing the impact of renewable energy integration on the systems strength [6].

1.3 The instantaneous generalised locus of impedances during power swing can also be used for tuning of relays depending on the intensity of the disturbance and swinging frequency.

2. Analytical derivation for the impedances

Table 1 below provides the nomenclature for the all symbols used in derivation.

Table 1. Nomenclature for symbols

Symbols	Nomenclature	Symbols	Nomenclature
V_p	Voltage at power frequency	I_p	Current at power frequency
V_s	Voltage at swinging frequency	I_s	Current at swinging frequency
V_{max}	Maximum peak voltage during swing	I_{max}	Maximum peak current during swing
V_{min}	Minimum peak voltage during swing	I_{min}	Minimum peak current during swing
v(t)	Voltage observed during power swing	i(t)	Current observed during power swing
ω_p	Power angular frequency	ω_s	Swinging angular frequency
m	Amplitude index	Φ	Phase shift at power frequency
k	Frequency index	Z_p	Impedance at power frequency
r(t)	Resistance during power swing	x(t)	Reactance during power swing

The voltages at power frequency and swinging frequency can be written in time domain as:

$$v_p(t) = V_p\cos(\omega_p t) \quad (1)$$

$$v_s(t) = V_s\cos(\omega_s t) \quad (2)$$

Power swing resembles an amplitude modulated waveform for both voltages and currents in time domain as shown in figure 1 with variations in amplitude and frequency [7]. The modulated wave in a AM signal is same as swinging wave during power swing and carrier wave in a AM signal is same as power frequency wave at 60Hz. The variation in differences of peak amplitudes depends on the intensity of the disturbance. Thus, higher the disturbance, greater is the differences between higher and lower peaks. The variation in swinging frequency depends on the system parameters like resistance, reactance, admittance and susceptance.

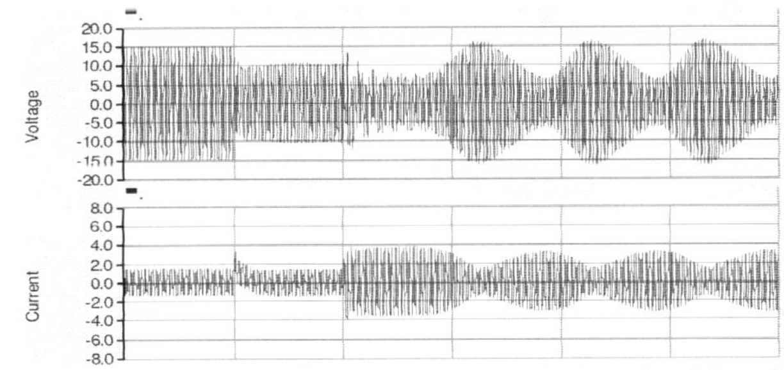

Figure 1. Power swing in voltages and currents. Reprinted from "A New Tracking Method of Symmetrical Fault during Power Swing Based on S-Transform" *(PEOCO2012)*, by N.Z. Mohamad et al. 6-7 June 2012 Malaysia [7]

Thus the amplitude modulated voltage [8] or a swinging voltage as a function of time can be represented as:

$$v(t) = [V_p + V_s\cos(\omega_s t)]\cos(\omega_p t) \quad (3)$$

Amplitude modulation index is defined as:

$$m = \frac{V_{max} - V_{min}}{V_{max} + V_{min}} = V_s/V_p \quad (4)$$

$$v(t) = V_p \cos(\omega_p t)\,[1 + m\cos(\omega_s t)] \quad (5)$$

Similarly, currents can be written with inclusion of delay angle in both power and swing frequency as:

$$i(t) = I_p \cos(\omega_p t - \phi)\,[1 + m\cos(\omega_s t - k\phi)] \quad (6)$$

m also indicates the ratio of maximum peak voltage to the minimum peak voltage or ratio of maximum peak current to the minimum peak current. It is same for both voltages and currents due to ohms law. If $V_s/I_s = V_p/I_p$ due to impedance; then $V_s/V_p = I_s/I_p$.

Finding impedance as a ratio of voltages and currents yields

$$z(t) = \frac{v(t)}{i(t)} = \frac{V_p \cos(\omega_p t)[1+m\cos(\omega_s t)]}{I_p \cos(\omega_p t-\phi)[1+m\cos(\omega_s t-k\phi)]} = Z_p \frac{\cos(\omega_p t)[1+m\cos(\omega_s t)]}{\cos(\omega_p t-\phi)[1+mco\ (\omega_s t-k\phi)]} \quad (7)$$

Rewriting the above equation in exponential format

$$z(t) = Z_p \frac{e^{j\omega_p t}[1+me^{j\omega_s t}]}{e^{j(\omega_p t-\phi)}[1+me^{j(\omega_s t-k\phi)}]} = Z_p e^{j\phi} \frac{[1+me^{j\omega_s t}]}{[1+me^{j\omega_s t}e^{-jk\phi}]} \quad (8)$$

Let x = $me^{j\omega_s t}$ in order to visually simplify the expression

$$z(t) = Z_p e^{j\phi} \frac{[1+x]}{[1+xe^{-jk\phi}]} = Z_p e^{j\phi}[1+x]\left[1+xe^{-jk\phi}\right]^{-1} \quad (9)$$

Binomially expanding the $(1+xe^{-jk\phi})^{-1}$ and rearranging the terms in ascending powers of x,

$$z(t) = Z_p\big[e^{j\phi} + xe^{j\phi} - xe^{-jk\phi}e^{j\phi} - x^2e^{-jk\phi}e^{j\phi} + x^2e^{-2jk\phi}e^{j\phi} + x^3e^{-2jk\phi}e^{j\phi} - x^3e^{-3jk\phi}e^{j\phi} - x^4e^{-3jk\phi}e^{j\phi} + x^4e^{-4jk\phi}e^{j\phi} + x^5e^{-4jk\phi}e^{j\phi} - x^5e^{-5jk\phi}e^{j\phi} - x^6e^{-5jk\phi}e^{j\phi} + x^6e^{-6jk\phi}e^{j\phi} \ldots\big] \quad (10)$$

Replacing x with $me^{j\omega_s t}$,

$$z(t) = Z_p\big[e^{j\phi} + m\big[e^{j(\omega_s t+\phi)} - e^{j(\omega_s t+\phi-k\phi)}\big] - m^2\big[e^{j(2\omega_s t+\phi-k\phi)} - e^{j(2\omega_s t+\phi-2k\phi)}\big] + m^3\big[e^{j(3\omega_s t+\phi-2k\phi)} - e^{j(3\omega_s t+\phi-3k\phi)}\big] - m^4\big[e^{j(4\omega_s t+\phi-3k\phi)} - e^{j(4\omega_s t+\phi-4k\phi)}\big] + m^5\big[e^{j(5\omega_s t+\phi-4k\phi)} - e^{j(5\omega_s t+\phi-5k\phi)}\big] - m^6\big[e^{j(6\omega_s t+\phi-5k\phi)} - e^{j(6\omega_s t+\phi-6k\phi)}\big] \ldots\big] \quad (11)$$

Eulers identity: $Ze^{j\phi} = Z\cos\phi + jZ\sin\phi = r(t) + jx(t)$

Separating the real and imaginary parts as resistance and reactance using Eulers Identity,

$$r(t) = Z_p[\cos\phi + m[\cos(\omega_s t+\phi) - \cos(\omega_s t+\phi-k\phi)] - m^2[\cos(2\omega_s t+\phi-k\phi) - \cos(2\omega_s t+\phi-2k\phi)] + m^3[\cos(3\omega_s t+\phi-2k\phi) - \cos(3\omega_s t+\phi-3k\phi)] - m^4[\cos(4\omega_s t+\phi-3k\phi) - \cos(4\omega_s t+\phi-4k\phi)] + m^5[\cos(5\omega_s t+\phi-4k\phi) - \cos(5\omega_s t+\phi-5k\phi)] - m^6[\cos(6\omega_s t+\phi-5k\phi) - \cos(6\omega_s t+\phi-6k\phi)] + \cdots] \quad (12)$$

Using cosine identities in ascending powers of m,

$$r(t) = Z_p\left[\cos\phi + 2\sin\left(\frac{k\phi}{2}\right)\left[-m\sin\left(\omega_s t+\phi-\frac{k\phi}{2}\right) + m^2\sin\left(2\omega_s t+\phi-\frac{3k\phi}{2}\right) - m^3\sin\left(3\omega_s t+\phi-\frac{5k\phi}{2}\right) + m^4\sin\left(4\omega_s t+\phi-\frac{7k\phi}{2}\right) - m^5\sin\left(5\omega_s t+\phi-\frac{9k\phi}{2}\right) + m^6\sin\left(6\omega_s t+\phi-\frac{11k\phi}{2}\right) + \cdots\right]\right] \quad (13)$$

Generalising the expression yields

$$r(t) = Z_p\left[\cos\phi + 2\sin\left(\frac{k\phi}{2}\right)\sum_{n=1}^{\infty}(-m)^n\sin\left(n\omega_s t + \phi\left(1-\frac{(2n-1)k}{2}\right)\right)\right] \quad (14)$$

Similarly, x(t) can be derived as

$$x(t) = Z_p\left[\sin\phi - 2\sin\left(\frac{k\phi}{2}\right)\sum_{n=1}^{\infty}(-m)^n\cos\left(n\omega_s t + \phi\left(1-\frac{(2n-1)k}{2}\right)\right)\right] \quad (15)$$

IOP Conf. Series: Materials Science and Engineering **605** (2019) 012010 doi:10.1088/1757-899X/605/1/012010

3. Test Bench using Real Time Digital Simulator

Real time digital simulator (RTDS) was used for testing the validity of the derived analytical equations in time domain which are r(t) and x(t). Figure 2 shows the test setup of the RTDS, RSCAD and computer workstation.

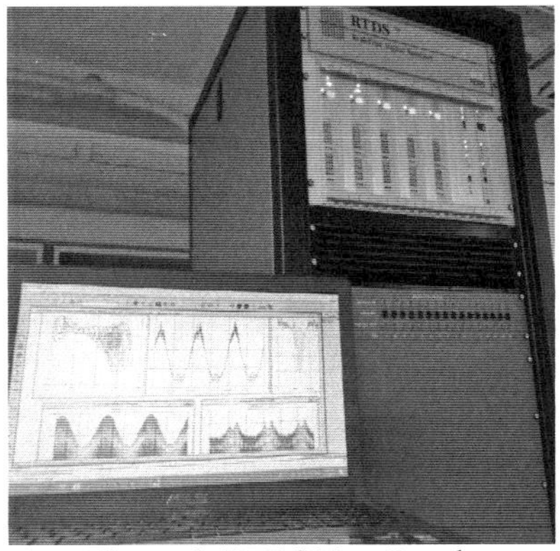

Figure 2. RTDS Test Bench

A simple 3 bus, 735kV, 2200MVA system was used consisting of generator, two transmission lines and a load. The measurements for voltages and currents were taken from the bus which is situated at the middle of the transmission lines. The generator model used can generate both the power frequency and swinging frequency. Transmission lines were modelled as long lines using toolbox from RSCAD. The load is a variable load which can be varied during run time. The r(t) and x(t) obtained will be plotted in two dimensional static impedance plot (R-X plane) and three dimensional spatial dynamic plot (R-X-t space) [9]. The test system used in RTDS is shown in the figure 3. The equations used for obtaining the impedances are shown below [10]:

$$Z(t)\angle\theta(t) = \frac{V(t)}{I(t)+dI_o(t)} = \frac{V(t)}{I(t)+d(\frac{Ia(t)+Ib(t)+Ic(t)}{3})} \quad (16)$$

where,

V and *I* are phase voltages and currents
d is the sensitivity factor for zero sequence current
Io is the zero sequence current.

$Z(t)\angle\theta(t)$ will be transformed to r(t)+jx(t) in order to compare the graphs with the derived equations.

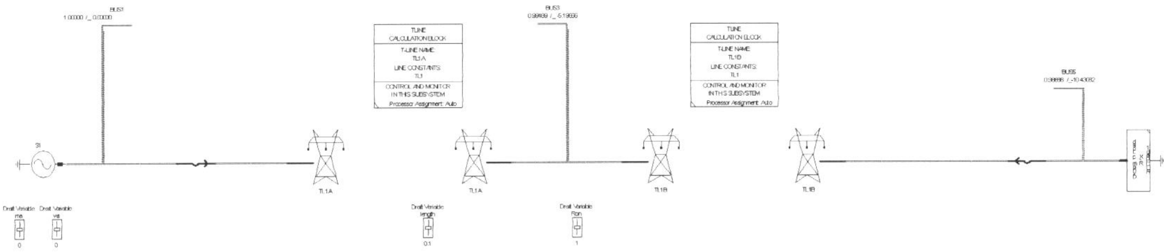

Figure 3. Circuit used in RTDS

4. Analytical Model vs RTDS

Validation of the analytical model using equations will be verified using RTDS. The validation will be done by varying power factor and swinging frequency for the analytically derived equations (14) and (15) and in RTDS test bench.

4.1. Variation of Power factor

Power factor will be varied from -60° to +60° for both the analytical model and RTDS. The power factors, reactive power and real power used are as shown in the Table 2.

Table 2. Different loading for 2200MVA system

Load	Power Factor Angle	Real Power	Reactive Power	Power Factor
2200 MVA	-60 °	1100 MW	-1905 MVAR	0.5 leading
2200 MVA	-30 °	1905 MW	-1100 MVAR	√3/2 leading
2200 MVA	0 °	2200 MW	0 MVAR	1
2200 MVA	30 °	1905 MW	-1100 MVAR	√3/2 lagging
2200 MVA	60 °	1100 MW	-1905 MVAR	0.5 lagging

All the simulations were carried out at the swinging frequency of 3Hz for different power factors.

4.1.1. Analytical Model

Power factor was varied for the analytical model according to the Table 2 using the equations (14) and (15). The R-X and R-X-t plots for the same are shown in the figures 4 and 5 respectively.

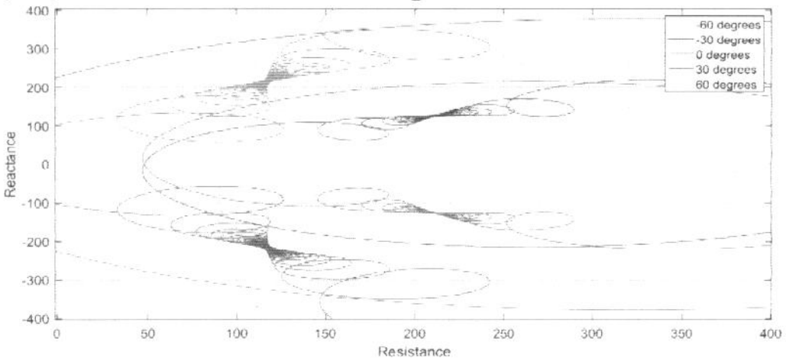

Figure 4. R-X using (14) and (15)

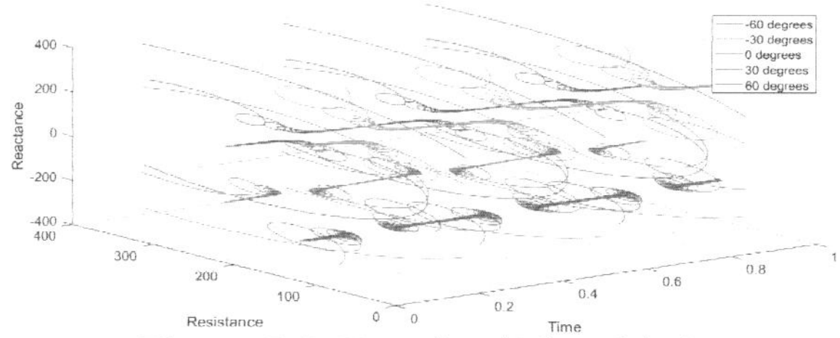

Figure 5. R-X-t using (14) and (15)

4.1.2. Real Time Digital Simulator

Power factor was varied for the RTDS variable load according to the Table 2. The two dimensional R-X and three dimensional R-X-t plots for the same are shown in the figure 6 and 7 respectively.

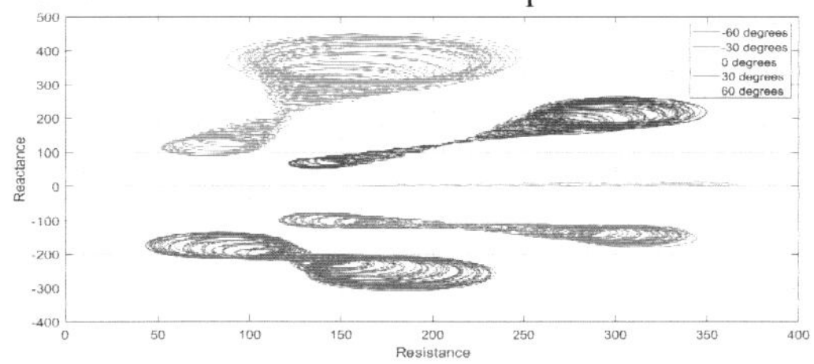

Figure 6. R-X plot using RTDS

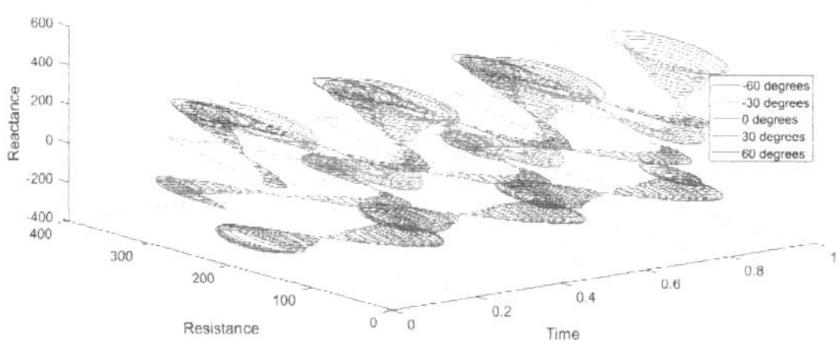

Figure 7. R-X-t plot using RTDS

4.2. Variation of Swinging frequency

Swinging frequency will be varied as 10/3Hz, 20/3Hz and 10Hz for both the analytical model and RTDS. This considers both slow swing which is 3.33Hz and fast swing which is 10Hz [11]. All the simulations were investigated at a power factor of 0.95 lead for different swinging frequencies.

4.2.1. Analytical Model

Swinging frequency was varied for the analytical model using the equations (14) and (15). The static R-X plot and dynamic R-X-t plots for the same are shown in the figures 8 and 9 respectively.

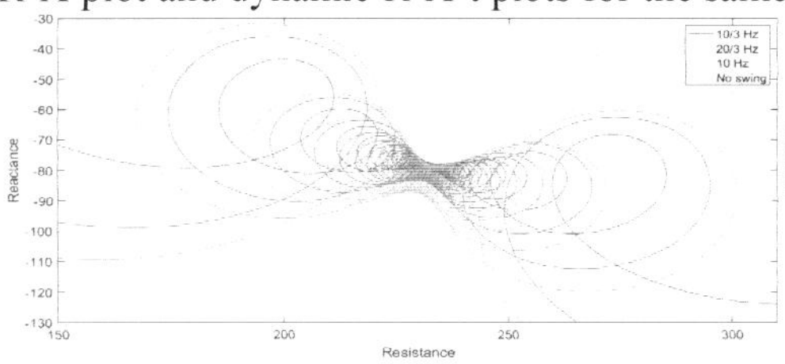

Figure 8. R-X using (14) and (15)

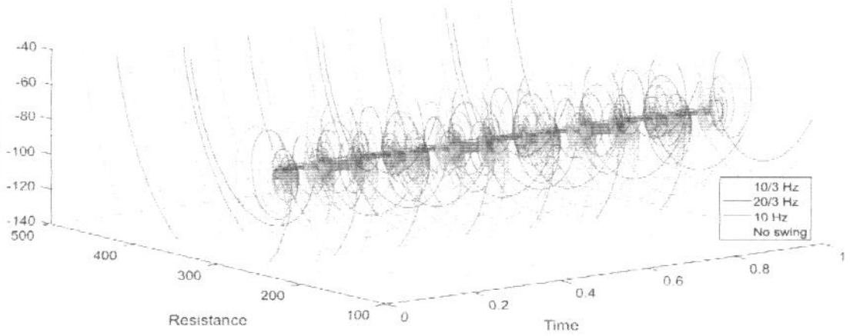

Figure 9. R-X-t using (14) and (15)

4.2.2. Real Time Digital Simulator

Swinging frequency was varied for the RTDS variable generator. The two dimensional R-X planar plot and three dimensional R-X-t spatial plots are shown in the figures 10 and 11 respectively.

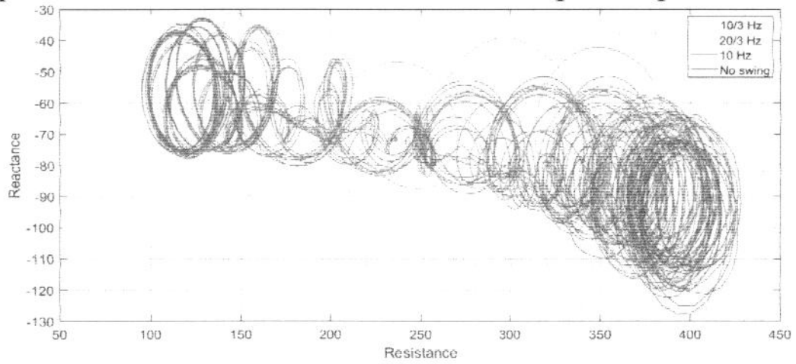

Figure 10. R-X plot using RTDS

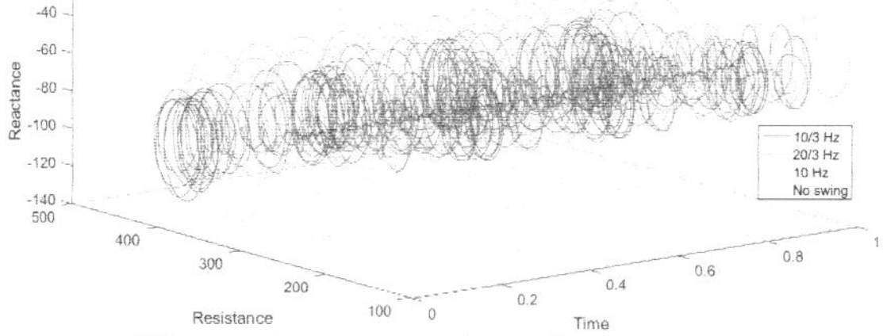

Figure 11. R-X-t plot using RTDS

5. Observations

The analytically derived equations and the plots taken from RTDS possess similarities and differences.

5.1. Variation in power factor

Figures 4, 5, 6 and 7 provide plots for variation in power factor according to Table 2.

Static two dimensional R-X plots for analytically derived equations and RTDS system are shown in figures 4 and 6 respectively. The major similarities observed in these two plots are the formation of biconical structures. The apex of the biconical structure is the normal operating point of the system in the absence of power swing (Zp = Vp/Ip). The circular formation extending beyond the normal operating point is due to the power swing. These are the swings responsible for unwanted tripping of relays. Both analytical expression and RTDS indicate that the swinging impedance undergoes phase shift depending on power factor and the oscillations being higher for lower power factors. At unity power factor, the plot is not visible in figure 4 but appears as a straight line in figure 6. This could be because the analytical equations did not take into account of any disturbances or noise which could appear in a real time digital power system.

Dynamic three dimensional R-X-t plots for analytically derived equations and RTDS system are shown in figures 5 and 7 respectively. Formations of biconical structures are evident in both plots. Figure 5 has connectivity between the outermost part of one cone to that of other cone apart from apex. This was not observed in figure 7, where the only connection between the two cones is the apex. Thus, apart from the outermost connecting loops, the inner loops are similar in both the analytical and RTDS plots. At unity power factor, figure 5 has a straight line passing along the normal operating impedance Zp indicating the absence of time varying resistance and reactance. In figure 7 there is an oscillating resistance which varies with respect to time which was not observed in figure 5. Thus, finer tuning is needed for the analytical expressions in order to accurately describe the locus of power swing as in real time.

5.2. Variation in swinging frequency

Swinging frequency was varied for both analytical and RTDS in steps of 3.333Hz as: 0Hz, 3.33Hz, 6.66Hz and 10Hz. Figures 8, 9, 10 and 11 provides plots during changes in swinging frequency.

Static two dimensional R-X plots for analytically derived equations and RTDS system are shown in figures 8 and 10 respectively. Though both plots cannot be completely superimposed, there are some major similarities in both the plots. The amplitude of swinging of impedance is proportional to the swinging frequency. It can be noticed that the 3.33Hz plot is inside the 6.66Hz which is inside 10Hz plot. Also, the number of oscillations in a specific duration was found to be low for 3.33Hz and high for 10Hz. In the absence of power swing, figure 10 indicates a small circle around the coordinates (235,-72). This explains the absence of any plot in figure 8 as it depicts a single accurate point due to the usage of analytically derived equation which is devoid of any uncertainty or disturbance.

Dynamic three dimensional R-X-t plots for analytically derived equations and RTDS system are shown in figures 9 and 11 respectively. The plot for absence of power swing was not clear in two dimensional plots, but they are clear in R-X-t plots. The absence of power swing in figure 9 was represented by a straight line parallel to time axis indicating the absence of any changes. In figure 11, absence of power swing was represented by a line which has very small oscillations with respect to R and X. Unlike analytical plots depicted by figure 9, the real time plots possess other minor unknown disturbances which could interfere with the measurements.

6. Conclusion

The analytically derived equation tends to approximately define the behaviour of power swing in time domain. Both the planar and spatial plots observed for the analytically derived equations and RTDS have some agreement with each other. Since measurements are taken at random from the transmission line, it will cover the disturbances caused by generators and loads. Thus grid disturbances happening within the centralised grid or integrated renewable microgrids will influence the r(t) and x(t) accordingly. The analytically derived equations require further tuning in order to fully track the locus of power swing in real time for all situations. This will later help in setting of relays only based on bus voltage, operating current and power factor without the requirement of sophisticated tooling or software for both centralised grids and integrated microgrids.

7. Future Work

The above derived analytical equations will undergo further refinement in order to completely fit the real time power swing. These parameters possess two applications as follows:

1. θ is tangent of X/R which is a measure of short circuit ratio. Analysing $\theta(t)$, derived from x(t) and r(t) will indirectly indicate the stiffness and immunity of the grid which is either entirely made of renewables or possessing higher percentage of renewables.
2. dr(t)/dt, dx(t)/dt and dθ(t)/dt will help in developing a robust technique for differentiating a fault, stable swing and unstable swing only based on time domain voltages and currents which can be easily obtained from measuring devices thus eliminating complicated numerical calculations.

8. References

[1] L. Lai, H.T. Zhang, C.S. Lai, "Investigation on July 2012 Indian Blackout", *International Conference on Machine Learning and Cybernetics*, July 2013, 978-1-47990260-6.

[2] A. S. Anees, "Grid integration of renewable energy sources: Challenges, issues and possible solutions", *IEEE 5th India International Conference on Power Electronics (IICPE)* 10.1109/IICPE.2012.6450514, 2012 Delhi, India

[3] S. Torres, H. Esponda, M. Andrade "Unblocking Function of Distance Relay during Power Swing Based on Modal Analysis" *IEEE PES Transmission & Distribution Conference and Exposition – Latin America (PES T&D-LA)*. Morelia, Mexico, ISBN 978-1-5090-2875- 7/16/©2016 IEEE.

[4] C. Pang, M. Kezunovic, "Fast Distance Relay Scheme for Detecting Symmetrical Fault During Power Swing" *IEEE Transactions On Power Delivery*, Vol. 25, No. 4, October 2010.

[5] X.Lin, Z. Li, H. Weng, "An Enhanced Power Swing Simulation Model Applicable to Protective Relaying Operation Behavior Analysis Based on EMTP Software", *Proceedings of the 2010 International Conference on Modelling, Identification and Control*, Okayama, Japan, July 17-19,2010

[6] D. Wu, G. Li, M. Javadi, "Assessing Impact of Renewable Energy Integration on System Strength Using Site-Dependent Short Circuit Ratio" *IEEE Transactions on Sustainable Energy*, Volume: 9, Issue: 3, pp 1072 – 1080, July 2018, DOI: 10.1109/TSTE.2017.2764871

[7] N.Z. Mohamad, A. F. Abidin, W. N. Munim,"A New Tracking Method of Symmetrical Fault during Power Swing Based on S-Transform" *2012 IEEE International Power Engineering and Optimization Conference (PEOCO2012)*,978-1-4673-0662-1/12/©2012 IEEE Melaka, Malaysia: 6-7 June 2012

[8] R.A. Heising "Modulation in Radio Telephony" *Proceedings of the Institute of Radio Engineers* Volume:9, Issue: 4, Aug. 1921 pp 305–352 DOI:10.1109/JRPROC.1921.220132.

[9] A. Zeno, J.R. Orillaza, "Discriminating Faults from Power Swing Using Impedance Angle and Analysis in Three Dimensional Space", *International Journal of Computer and Electrical Engineering(IJCEE)* Volume: 11(1), pp27-40, March 2019, ISSN: 1793-8163

[10] Y.G. Paithankar, S.R. Bhide, *Fundamentals of Power Systems Protection*, Prentice Hall, 2011, p145.

[11] Milad Afzali, Ahad Esmaeilian "A Novel Algorithm to Identify Power Swing Based on Superimposed Measurements" 978-1-4577-1829-8/12/©2012 IEEE

Comparative analysis of Artificial Neural Networks with conventional methods for extrapolation of wind speed at an elevated height

Dhrumin Bhavsar[1*], Mohil Bhatt[1], Mehul Choksi[1] and Dr. Garlapati Nagababu[1]

[1] Department of Mechanical Engineering, PDPU, Gandhinagar-382007, Gujarat, India

[*]dhrumin2203@gmail.com

Abstract. Due to rapidly increasing pollution, it becomes necessary to substitute fossil fuels, and as wind energy is available quite easily and in abundance, researches are carried out in this area. These facts make it imperative to know about the variables and the problems involved behind it. The wind speed is a random variable, and it depends on atmospheric factors like pressure, relative humidity, wind dispersion & wind direction. This paper introduces the method to effectively predict wind speed by making use of the Levenberg-Marquardt backpropagation algorithm in artificial neural network (ANN) and by conventional means like power law & log law in MATLAB. Data from the Gulf of Khambhat, Gujarat provided by LIDAR for a period of 8 months, was used to prepare valid data set to train the neural network and to build a model to predict wind speed. After obtaining a histogram of the predicted values by log law, power law and ANN, it was seen that wind speed values obtained by ANN were quite close to actual values than the values obtained through the other methods. A comparison in terms of root means square error and percentage of the number of data indicates that developed neural network gives less root mean square error and a higher percentage of data whose absolute error lie between -0.2 and 0.2.

1. Introduction

As the population and standard of living of human increases, the energy consumption rises and the need for energy expands rapidly. If we continue to produce energy in the same conventional ways, then we are in turn contaminating the environment. So in order to meet the increased energy demands and environmental factor we have to have some alternative energy sources. Among the renewable energy resources, wind energy is a more feasible and easily available energy source, and it is also cheaper than solar energy resource [10]. The target of 15 percent of renewable power by 2020 can be achieved only if the wind energy sector is allowed to grow [8]. The wind turbine design is dependent on the wind speed at the turbine hub height [11]. Overdesigning the turbine blade which leads to an increase in the size of the blade is a big issue in terms of financial investment [7]. So to design particular wind turbines accurate wind speed is required at the turbine hub height. Moreover, in many cases measuring the wind speed at elevated height is difficult. So wind speed needs to be extrapolated at elevated height by taking reference of wind speed at a lower height. Accurate extrapolation of the wind speed over different elevation is a complex problem. To extrapolate the wind speed at elevated heights, generally, two conventional methods are used. One is extrapolation using power law and the second one is extrapolation using log law. These methods depend upon constant exponents, which does not consider the change in environmental condition and other change in wind-related parameters which are directly or indirectly affects the wind speed at different heights. At lower altitudes wind speed follows the logarithmic trend but when the data of surface roughness is not available, then it is challenging to predict the wind speed accurately with the log-law approach. Wind shear coefficient is dependent on environmental stability and wind characteristics [9], which states that wind speed is also dependent on those parameters. Neural networks approach can consider environmental factors on which wind speed is dependent. A Neural Network can develop an algorithm which can predict accurate outputs by initially training the network with the known data in which dependent parameters are taken into consideration. ANN can learn and adapt to complex and nonlinear data, and most importantly ANN does not impose any restriction on input data variables. In this study a Feedforward Neural network is developed using the 8-month dataset at 40m, 60m, 80m and 87m heights at 10 minutes time interval by considering six input parameters. LIDAR collects these data at the location of Gulf of Khambhat, Gujarat, India. In this study wind speed is extrapolated over 100m to 200m height thorough three different extrapolation method and results have been compared in terms of the root mean square errors and percentage of the given data for which absolute errors lies within the range of [-0.2,0.2].

Nomenclature

v_{ext}	Extrapolated Velocity	m/s	w	Weight
v_{ref}	Velocity at the reference height	m/s	B	bias
z_{ext}	Extrapolation Height	m	y_n	Predicted output using Neural Network
z_{ref}	Reference Height	m	n	Index of output data
e	Absolute Error	m/s	k	Index of Iteration
z_o	Roughness Length	mm	E	Absolute error Matrix
v_{tr}	Training Dataset Velocity	m/s	N	Total output data
z_{tr}	Training Dataset height	m	J	Jacobian Matrix (Partial Derivative of the Error function with weights and bias)
MSE	Mean Square Error	$(m/s)^2$	α	Wind Shear Coefficient
RMSE	Root Mean Square Error	m/s		
v	Actual Velocity of the dataset	m/s		

2. Collected Data

The input variable plays a vital role to achieve accurate results. To estimate the wind speed accurately study of the effect of the different input parameter is essential. The LIDAR based measurement was conducted in Gulf of Khambhat, Gujarat, India by National institute of Wind Energy Chennai and wind related data in 10 minute time interval were collected in time span of November 2017 to June 2018. The data consist of parameters such as Pressure, Relative Humidity, Height, Wind dispersion, Wind Direction and wind speed from November 2017 to June 2018. LIDAR collects these data at the location of Gulf of Khambhat, Gujarat, India. The data is of total eight-month duration and at the interval of 10 minutes. Wind speed varies from season to season, and even from day to night, it is because of the change in the temperature and pressure gradient. As the difference gets higher, the speed of the wind also increases [12].

3. Methodology

The Speed of height ranging from 100m to 200m are extrapolated using two conventional methods and also with a developed feed-forward neural network approach. These extrapolated Speed values are compared with the actual ones, and overall RMSE is obtained. After that RMSE value of all three methods is compared to conclude which method is best suitable for the given problem.

3.1. Power Law

Power law method is one of the most common and conventional methods used to extrapolate wind speed over different heights. Here it is used to extrapolate wind Speed from 100m to 200m height. The Speed is extrapolated over a dataset of 8 months with 10 minute time interval at 100m, 107m, 120m, 140m, 160m, 180m and 200m using power law.

As a reference, 8-month data at 10 minutes time interval of 40m is taken into consideration to extrapolate wind speed at different heights. The wind shear coefficient (α) is taken as 1/7. The following equation is used for Extrapolation of Speed from 100m to 200m height [1].

$$v_{ext} = v_{ref} \times \left(\frac{z_{ext}}{z_{ref}}\right)^{\alpha} \tag{1}$$

3.2. Log Law

Log law method is another commonly used conventional method to extrapolate wind speed over different heights. The Speed is extrapolated over the same dataset of 8-months with 10 minute time interval at 100m, 107m, 120m, 140m, 160m, 180m and 200m with log law.

Corresponding 8-month data at 10 minutes time interval of 40m is considered to extrapolate the wind speed over different elevated height. As per the NIWE report of LIDAR wind resource assessment, the average value of surface roughness is taken as 0.2 [12]. The following equation is used for Extrapolation of wind speed over 100m to 200m height [1].

$$v_{ext} = v_{ref} \times \frac{\ln\left(\frac{z_{ext}}{z_o}\right)}{\ln\left(\frac{z_{ref}}{z_o}\right)} \tag{2}$$

3.3. Neural Network

The main objective of using a neural network approach is to improve the performance of extrapolated wind speed by taking factors like time horizon, two environmental conditions (Temperature and pressure), two wind-related parameters (Wind dispersion and wind direction) and height into consideration. A feed-forward neural network was created and trained using Levenberg-Marquardt backpropagation algorithm considering eight-month training data of 60m, 70m, 80m and 87m with six input variables and one output variable as wind shear coefficient. The developed feed forward neural network is used to predict wind shear coefficient for 100m, 107m, 120m, 140m, 160m, 180m, and 200m height. The predicted wind shear values are used to calculate the wind speed at 100m, 107m, 120m, 140m, 160m, 180m, and 200m altitude

3.3.1. Feed Forward Network. A feed-forward neural network consists of six input layer neurons, eight hidden layer neurons, and one output layer neuron. Six input layer neurons are taken since we have six desired input variables, which are Time Horizon, Pressure, Relative Humidity, Height, Wind dispersion and Wind Direction. One output neuron is considered for wind shear coefficient as it is the only output variable. After many iterations for training the networks, the decision to take eight number of hidden layer neurons was finalized to develop neural network optimally.

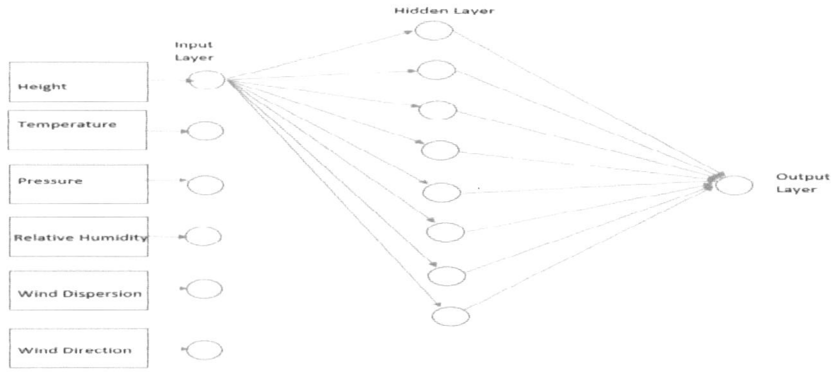

Figure 1. Schematic Diagram of Feed Forward Network

For representation purpose only one input neuron connections with all hidden neurons are shown in Figure 1.

Actual network has all input and hidden neurons are connected in forward direction.

Table 1. Feed Forward Neural Network Selected Parameters

Neural Network Parameter	Parameter value/function
Number of Input Layer Neurons	6
Number of Input Layer Neurons	8
Number of Input Layer Neurons	1
Hidden layer transfer function	tansigmoid
Hidden layer transfer function	purelinear

3.3.2. Preparing a training target. Wind shear coefficient is taken as an output variable of a feed-forward neural network. For training the data, the exact values of wind shear coefficient are required. These values are calculated for an eight-month training dataset of 60m, 70m, 80m, 87m at 10 minutes time interval by keeping α as a subject in power law equation [7]. Values of α are calculated for all N number of data available at each height.

$$\alpha = \frac{\ln v_{tr} - \ln v_{ref}}{\ln z_{tr} - \ln z_{ref}} \quad (3)$$

3.3.3. Training Neural Network. After creating a feed-forward neural network an eight-month training dataset of 60m, 70m, 80m, 87m at 10 minutes time interval was taken to train and validate the feed-forward neural network. The training algorithm used was the Levenberg-Marquardt backpropagation algorithm.

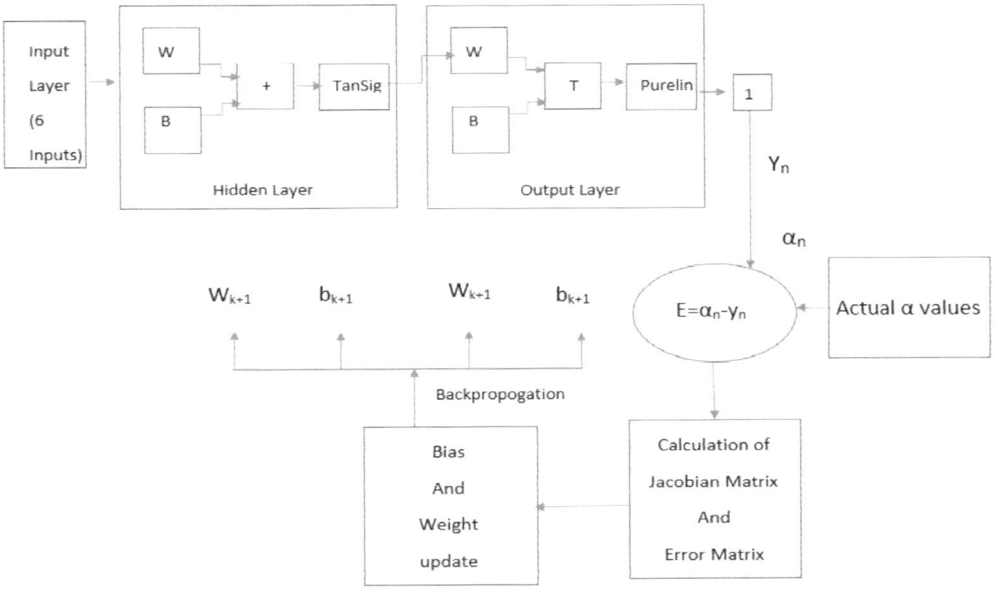

Figure 2. Schematic Diagram of Training phase

Training starts with randomly initialization of weight and bias between neurons of input layer-hidden layer and Hidden layer-Output layer. In the first iteration taking into consideration of all input variable values multiplies with corresponding weight and adding each bias of each neurons gets added and gives output to the upcoming neuron through assigned transfer function. During this only forward propagation a certain output at output neuron is generated. This output is compared to the actual values of training dataset in terms of error function MSE [5]. Than during back propagation at each neuron absolute error and partial derivatives of absolute error with respect to weights and bias are calculated, forming jacobian matrix (J) and absolute error matrix (E). And the new weights and bias at each neurons are updated using following equation [2]. In each kth iteration jacobian matrix and absolute error matrix gets updated. Combination Coefficient is taken as 0.1.

$$e_n = \alpha_n - y_n \tag{4}$$

$$MSE = \frac{\sum_{n=1}^{N} {e_n}^2}{N} \tag{5}$$

$$w_{k+1} = w_k - [{J_k}^T J + \mu I]^{-1} {J_k}^T E_k \tag{6}$$

3.3.4. Calculation of extrapolated wind speed. The trained neural network is used to predict the wind shear coefficient (α) by feeding the input parameters of a dataset consisting of 100m, 107m, 120m, 140m, 160m, 180m, and 200m height. The predicted α values are fed into the power law equation to extrapolate wind speed over these heights.

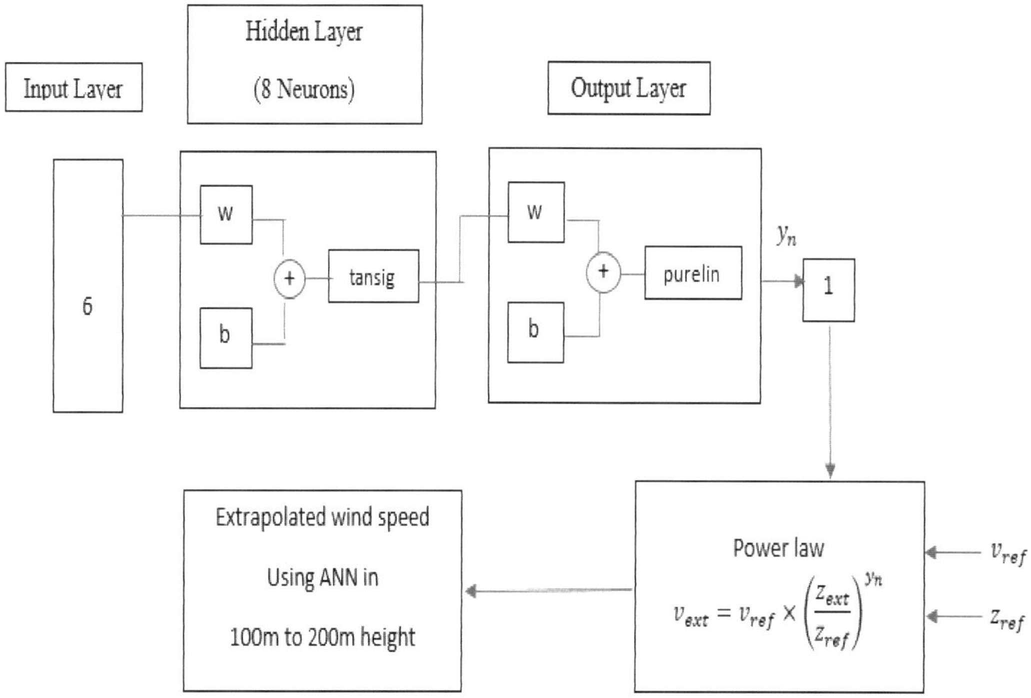

Figure 3. Schematic Diagram of Extrapolation of wind speed using trained Neural Network

3.4. Error Comparison
RMSE calculates the square root of the averaged squared errors between actual output and extrapolated output by a particular method, and it is more useful for more significant errors [5]. So among the standard comparison methods, RMSE and the absolute error is selected to compare three different extrapolation methods. After extrapolating wind speed of eight-month dataset at 100m, 107m, 120m, 140m, 160m, 180m and 200m height using power law, log law and developed Feed-forward neural network, at each height for N data an absolute error and RMSE [5] between extrapolated wind speed and actual wind speed were calculated.

$$e_n = v_n - y_n \tag{7}$$

$$RMSE = \sqrt{\frac{\sum_{n=1}^{N} {e_n}^2}{N}} \tag{8}$$

4. Results
To compare the performance of three extrapolation methods two comparison parameter was used. First one is RMSE and the second comparison parameter is the percentage of the given data for which absolute error lies in the range of [-0.2, 0.2].

4.1. Result containing RMSE
RMSE gives information about the overall error of specific data; that is why it gives a better understanding of the comparison of the performance of different methods [5]. After extrapolating wind

speed of eight-month dataset at 100m, 107m, 120m, 140m, 160m, 180m and 200m height using power law, log law using equation (1) and (2) respectively, at each height for given data the actual wind speed and extrapolated wind speed was given to the equation (8), and RMSE value was obtained between extrapolated wind speed and actual wind speed for N data at each height.

After finding RMSE using two conventional methods, a trained neural network was used to predict α at 100m, 107m, 120m, 140m, 160m, 180m, and 200m height. Extrapolated wind speed was calculated Using equation (1) at each height. The extrapolated wind speed and Actual wind speed are put into equation (8) to get RMSE at each height 100m, 107m, 120m, 140m, 160m, 180m, and 200m. Calculated RMSE at each height using three various extrapolation methods is shown in Table 2. The RMSE at 100m, 107m, 120m, 140m, 160m, 180m and 200m for three extrapolation methods are also represented graphically in Figure 4.

Table 2. RMSE for Extrapolation of wind speed using three different methods at various height.

Training Data in Neural Network	Extrapolated Height in Neural Network (m)	Root Mean Square Error (m/s)		
		Log Law	Power Law	Feed Forward Neural Network
60m to 87 m	100	0.989	1.041	0.936
	107	1.078	1.155	1.006
	120	1.232	1.273	1.153
	140	1.44	1.522	1.359
	160	1.639	1.732	1.553
	180	1.817	1.938	1.734
	200	1.968	2.070	1.878

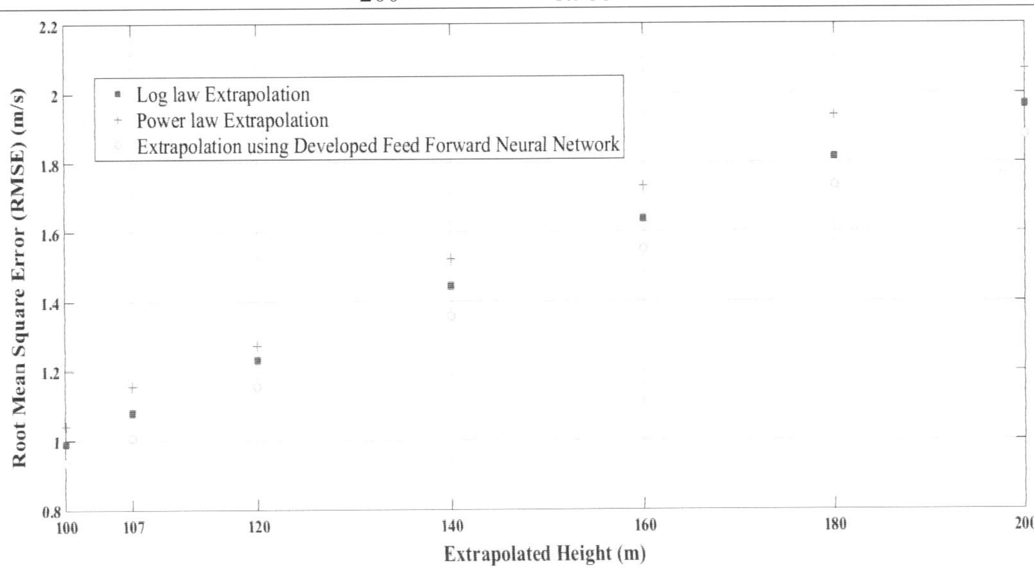

Figure 4. Scatter plot of RMSE (m/s) in all extrapolation methods.

From Table 2. And Figure 4., it was observed that at every height the overall RMSE for extrapolated velocities over various heights is Highest for conventional power law method while the feedforward neural network method gives least overall RMSE at all heights. Log law has slight higher RMSE error value compared to neural network approach, but it is better than the power law approach for extrapolation. Power law seems to give the worst result of RMSE compared to other two extrapolation method approach. It was observed that the performance of all three methods decreases with an increase in height

4.2. Results containing % Data with error in the range [-0.2, 0.2]

It is also essential to find the absolute error between extrapolated values and actual values. A method whose absolute errors are near to zero is expected to give more accurate results. An absolute error has been calculated using equation (7) just by taking the difference between extrapolated and actual wind speeds for each height.

Similarly after finding absolute error using two conventional methods a trained neural network has been used to predict α at 100m, 107m, 120m, 140m, 160m, 180m, and 200m height. Using equation (1) extrapolated wind speeds were calculated at each height. The extrapolated wind speed and Actual wind speed are fed into equation (7) to get the absolute error (e) at each height 100m, 107m, 120m, 140m, 160m, 180m, and 200m. Calculated absolute errors of all heights combined are represented graphically as a histogram using three various extrapolation methods.

Also since it is essential to know how much-extrapolated data are accurate (error within [-0.2, 0.2]) percentage of data with error within [-0.2, 0.2] is also calculated and represented in all three histograms with three different extrapolation method. Percentage of data with error within [-0.2, 0.2] for each technique is also described in Table 3.

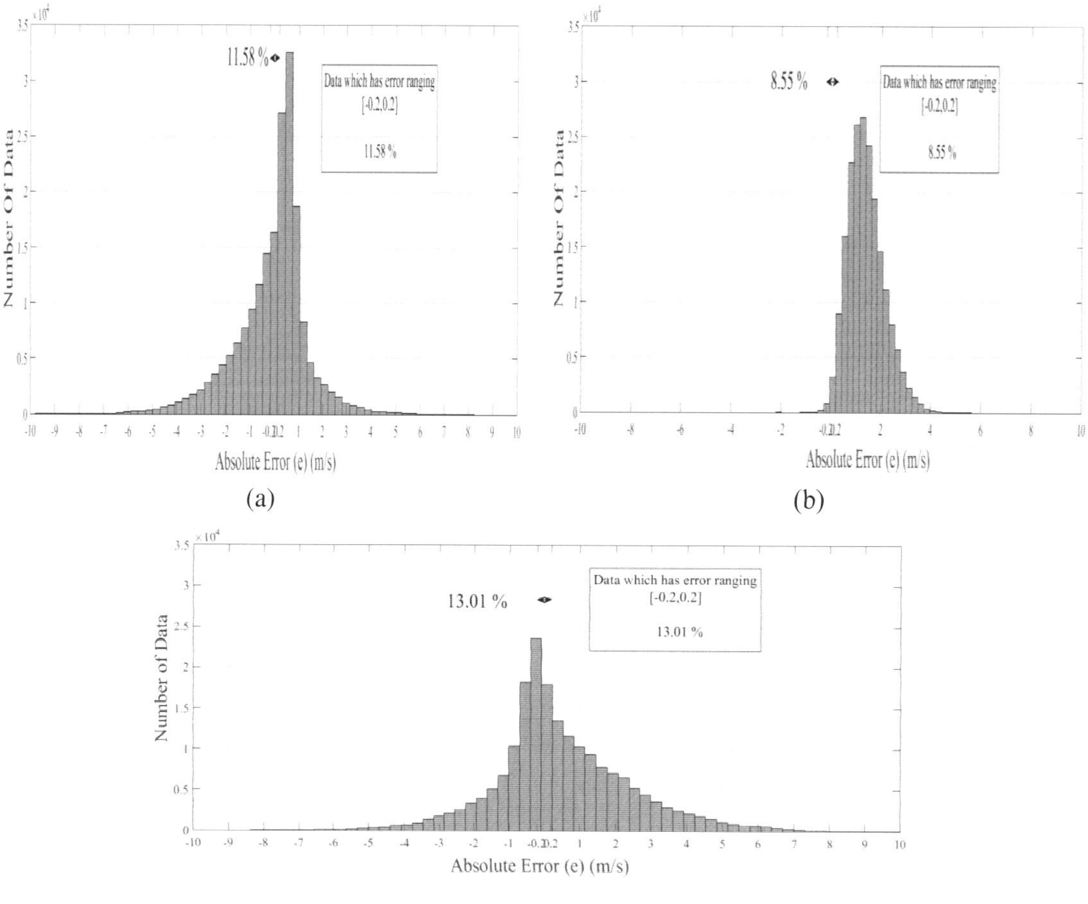

(c)

Figure 5. Absolute error histogram of Extrapolated wind speed using (a) Log Law (b) Power Law (c) Feed Forward Neural Network

Table 3. % Data whose absolute values (m/s) are between -0.2 and 0.2

Log Law	Power Law	Feed Forward Neural Network
11.58	8.55	13.01

From Figure 5(a)., Figure 5(b)., Figure 5(c). And Table 3. , it can be observed that percentage data with extrapolated wind speed absolute error (m/s) within [-0.2, 0.2] is highest (13.01) in Developed Neural Network. And least in Power law (8.55). From all three histogram, histogram of Developed Neural Network seems more converged towards zero error.

Log law method histograms seems converged towards error in range of [0.7, 1] which is slight bad than neural network approach histogram. And also it has 11.58% of data within absolute error range [-0.2, 0.2]. Which is better than the power law bus litter worse than the neural network approach.

Power law histogram shows the worst result as the absolute error (m/s) within [-0.2, 0.2] is least (8.55), and the histogram seems more converged towards error range [0.7, 1.5].

5. Conclusion

From the study and observation of results obtained of comparisons between three extrapolation methods in terms of RMSE and the percentage number of data in the range of [-0.2,0.2] following conclusions has been drawn.

- Extrapolation of wind speed using trained feedforward neural network shows least overall RMSE over two other conventional methods.
- Extrapolation of wind speed using trained feedforward neural network gives the maximum number of data (13.01%) whose absolute error is in the range of [-0.2,0.2] compared to two other conventional methods.
- Since conventional methods does not consider changes in environmental and changes in wind parameters, they give lower performance than the ANN approach for extrapolation of wind speed at elevated height.
- Since overall results of Neural Network approach is better than the conventional methods, we can say that since Neural network approach considers changes in some parameters which directly or indirectly are related to the wind speed at different heights and different time horizons, Neural network approach gives better result than the conventional methods.
- Since in ANN approach the data used for training of network is only for one particular location, ANN approach is limited for extrapolation of wind speed at the location of Gulf of Khambhat, Gujarat only. While conventional methods are applicable to any location resulting in low extrapolation performance.

6. References

[1] Dr. Firas A. Hadi "Diagnosis of the Best Method for Wind Speed Extrapolation," International Journal of Advanced Research in Electrical, Electronics and Instrumentation Engineering, Vol. 4, Issue 10, October 2015.
[2] Yu, Hao & Wilamowski, Bogdan. (2011). Hao Yu and B. M. Wilamowski, Levenberg Marquardt Training Industrial Electronics Handbook, vol. 5 Intelligent Systems, 2nd Edition, chapter 12, pp. 12-1 to 12-15, CRC Press 2011...
[3] Vehbi Olgac, A & Karlik, Bekir. (2011). Performance Analysis of Various Activation Functions in Generalized MLP Architectures of Neural Networks. International Journal of Artificial Intelligence And Expert Systems. 1. 111-122.
[4] A. Khosravi, R.N.N. Koury, L. Machado, J.J.G. Pabon, "Prediction of wind speed and wind direction using artificial neural network, support vector regression and adaptive neuro-fuzzy inference system" Sustainable Energy Technologies and Assessments 146–160, 2018.

[5] Jamil, Majid, and MohammadAli Zeeshan. "A comparative analysis of ANN and chaotic approach-based windspeed prediction in India."Neural Computing and Applications(2018)1-13.

[6] Nand Kumar, Gurpreet Kaur, Aditim," Wind Speed Prediction using Neural Network," International Journal of Advanced Production and Industrial Engineering IJAPIE-SI-IDCM 608 36–41.2017

[7] Nawal Cheggaga and Fatiha Youcef Ettoumi, "A Neural Network Solution for Extrapolation of Wind Speeds at Heights Ranging for Improving the Estimation of Wind Producible" Wind Engineering 35(1):33-54, 2011.

[8] Deepika Chauhan, Sanjay Agarwal, Mahendra Kumar Suman, "Wind Power scenario in India: A Review", International Journal of Scientific and Research Publications, Volume 3, Issue 9, September 2013 ISSN 2250-3153

[9] John S. Irwin, "A theoretical variation of the wind profile power-law exponent as a function of surface roughness and stability", Atmospheric Environment (1967), Volume 13, Issue 1,1979, Pages 191-194, ISSN 0004-6981

[10] Kasra Mohammadi, Ali Mostafaeipour, Majid Sabzpooshani, "Assessment of solar and wind energy potentials for three free economic and industrial zones of Iran", Energy, Volume 67,2014, Pages 117-128, ISSN 0360-5442

[11] Tavner, P.J. & Edwards, Clare & Brinkman, Andy & Spinato, Fabio. (2006), "Influence of Wind Speed on Wind Turbine Reliability", Wind Engineering. 30. 55-72. 10.1260/030952406777641

[12] First offshore Lidar Wind Data Analysis Report –national Institute of wind energy (NIWE), wind resource assessment & offshore unit Chennai, India,2018

Acknowledgment
We want to express our gratitude towards Pandit Deendayal Petroleum University for assisting with research from all possible means.

IOP Conf. Series: Materials Science and Engineering **605** (2019) 012012 doi:10.1088/1757-899X/605/1/012012

Reactive Power Control of a 3-phase Grid-Connected Building Integrated Photovoltaic System

M J M A Rasul and Mohan Kolhe

Dept. of Engineering Sciences, Faculty of Engineering and Science,

University of Agder (UiA), N-4898 Grimstad, Norway

ashikmar@gmail.com, mohan.l.kolhe@uia.no

Abstract. Recently, a tendency in the growing of grid-connected building integrated photovoltaic (BIPV) systems has been noticed in most countries. Hence, high penetration of PV power into the system network can be observed in many points in the network. This may cause severe problems in the distribution network originated by the intermittent behaviour of PV systems. As a feasible solution, the reactive power capability of voltage source inverter in PV systems can be employed rather than approaching expensive grid infrastructures in the distribution network. The purpose of this study is to implement a 3-phase grid-connected BIPV system with reactive power control to regulate the system voltage and improve the system power factor. Subsequent to this, the system model is designed and developed in MATLAB Simulink application. The inverter control system along with the reactive power control is developed for the proper system operation at different system conditions to reduce the var compensation on the utility grid. A better system accuracy of the developed system model is verified by validating with real system data.

1. Introduction

Presently, a tendency in the rapid growth of energy has been noticed in the world. Hence in most of the countries, renewable energy resources (RES) have been utilized in electricity production. The main available RESs are hydroelectric, biomass, photovoltaic, wind, geothermal and tidal systems and also called the alternative source of energy which can be used again and again. Consequently, this increase in renewable energy generation systems combined with power electronic systems, the microgrid concept has been developed. The capability of operating in an uninterruptible way either grid-connected or islanding mode is the main benefit of a microgrid.

Integrating RESs such as wind and solar, into the utility grid at the distribution level, is a trending fact. These distributed energy resources (DERs) supply power directly to the grid via power electronic converters where the final stage is a DC-AC inverter between the DERs and the grid. In recent in years, Photovoltaic (PV) system has been the fastest growing power generation system worldwide, has a total capacity of 228 GW at the end 2015 [1]. Generally, the available types of PV systems are residential scale, commercial scale and utility scale in which capacities range from few kilowatts to hundreds of megawatts.

These PV systems inject real power to the grid via at many points in the distribution network. Since this is an effective solution for rapid growing energy demand, but this will bring adverse consequences on the entire distribution network. This is because of the intermittent behavior of the PV energy which accordingly have an impact on the availability, reliability, and quality of the distributed grid. Due to this intermittent nature and cumulative real power injections at different points, it is a very much complex process of making reliable estimations of the power circulations throughout the distribution network employing a few readings at a few different locations. In case of high PV production level, the whole power system will be out of control and there is a potential risk of significant reverse power circulation, and accordingly causing unacceptable voltage hike in the network voltage [2]-[4].

To overcome this abnormal voltage over-limits situations, the expensive grid infrastructures such as under-load tap changing transformers (ULTC), step voltage regulators (SVR), and fixed and switchable capacitors (FC and SC), can be deployed or the reactive power control ability of voltage source inverter (VSI) in PV systems can be utilized for injecting or absorbing reactive power for voltage compensation as an ancillary service [5],[6]. The advantages of utilizing PV inverters over expensive grid infrastructures are the capability of providing continuous variable reactive power which enables to manage the voltage more closely, and inverter quick response in comparison with switched capacitors [7].

Previously the IEEE Standard for Interconnecting Distributed Resources with Electric Power Systems (IEEE Std 1547, 2003) [8] states that the DERs are not permitted to regulate the voltage at the point of common coupling (PCC). Due to some technical problems arise with high PV penetration level after rapid growth in PV energy production, the amendment IEEE Std 1547a-2014 [9] unleashes the restrictions that the DERs are allowed to actively engage in regulating the grid voltage by making alterations to the active and reactive power productions of the DERs coupled with mutual agreements with the distribution grid operators and the DERs.

Hence, the PV inverters can be utilized as the fast acting local var compensators in the areas with significant levels of PV penetration, providing reactive power absorbed by local loads as measured at their point of common coupling (PCC) and subsequently help to enhance the power factor (PF) at PCC. This will assist to attenuate the voltage regulation problems along with the PF improvement while enhancing the grid stability and maximizing the power transmission capability by reducing transmission loss. In the meantime, this will be a significant economic benefit for consumers to avoid the cost of the equipment required to control the PF.

With the rapid growth in the penetration level of PV power on the distribution system, voltage fluctuations issues can be arisen around the location of the interconnection of PV systems due to the real power injection of PV systems. The reactive power control functionality of PV inverter has been utilized to support the voltage regulation at the distribution level. In recent years, most of the researches have been focused on the reactive power control strategies for grid-connected PV inverters for active contribution on the voltage regulation.

Different techniques have been discussed for the voltage support, based on the reactive power compensation of the PV inverter in literature. Those can be grouped as fixed PF (fixed cosφ), PF as a function of the active power (cosφ(P)) fed into the network, fixed reactive power (fixed Q), reactive power as a function of the voltage at the point of common coupling (Q(V)) [10],[11] and reactive power as a function the active power (Q(P)) [12]. Besides these technologies, several studies have been presented based on the reactive power control along with the active power curtailment for voltage regulation problem [13]-[15].

Reference [16] presented a three-phase grid interactive PV inverter with reactive power ability to support the grid voltage regulation. The proposed system was enabled to operate in unity PF, variable PF, and zero PF to meet the grid requirements. In [17], an adaptive decentralized voltage restriction control by the autonomous reactive power control of grid-connected PV inverters was presented and tested on different LV grids to reduce the necessity of additional grid reinforcement measures. The proposed strategy based on the V-Q characteristics with a dead band where the reactive power injection or consumption by the inverter was initialized when the voltage limitations exceed. The work in [18] addressed a reactive power control (Q(V)) relying on voltage for an inverter with a PF of 0.9 for compensating the voltage rise due to the PV power feed in an MV network located in Belgium.

A power control approach based on the single-phase active-reactive power theory which was controlled by system conditions and specific demands from both system operators and customers was presented in [19] to enable the PV inverters to perform the multi-functional ancillary services such as "low voltage ride through (LVRT), reactive power compensation and reliability-oriented thermal management" on PV inverters. The prescribed power control strategy was developed and utilized under the limitation of the PV inverter ratings. Similarly, in [20], an inverter control strategy was based on the PQ theory which facilitates to produce the active power and compensate the reactive power of the loads during daytime while only compensating the reactive power of the loads during night time.

Utilizing analysis of voltage sensitivity by setting a location dependent PF to each PV inverter units for the grid voltage support control was presented in [21] as a decentralized reactive power. In this work, the combination of aforementioned strategies, cosφ(P) and Q(V), was employed to avoid the unnecessary absorption or injection of reactive power during over-voltage conditions. A similar approach using the voltage sensitivity analysis was discussed in [22] for the reactive power control of the PV systems, to attenuate the voltage fluctuations due to the active power variations of the PV systems. The effect on the inverter efficiency was discussed for three different modes of reactive power control to investigate the inverter performance.

In [23], an inverter reactive power injection control was designed, based on the volt/var control as a "radial optimal power flow (OPF) problem", subjected to the constraints of voltage and PV inverter reactive power limit where the objective was to reduce the line losses, the energy consumptions and the cost of inverter losses when not operating at unity PF. Reference [24] discussed a method, based on the absorption and injection of uneven reactive power by three-phase PV inverters, to avoid the voltage unbalance due to the high penetration of single-phase inverters in the LV grid. In this proposed method, the average values of the phase voltages at the PCC of the PV inverters were employed as the references for the balancing algorithm for controlling each phase independently to achieve uneven compensation and also reducing the over-voltage conditions.

An active power dependent (APD) (Q(P)) voltage regulation method was proposed in [25] in which the required reactive power to manage the voltage was calculated in terms of the penetrated real power of a PV system. In this proposed method, an active power threshold, based on the investigation of the "voltage sensitivity matrix", was employed as a start point above which the PV system was ready to consume reactive power to regulate the voltage. Reference [26] discussed an active power dependent reactive power compensation method without utilizing a threshold into the system where the aim was to optimally dispatch the inverter reactive power while maintaining the voltage within an admissible range and minimizing the resistive losses over the entire network. An "optimal inverter dispatch (OID)" framework was proposed in [27] where the task was to determine both the real and reactive power set points of the PV inverter so that to optimize the network operation by minimizing the power losses while assuring the voltage regulation and adhering to other electrical network constraints. The proposed OID framework increased the adaptability over the volt/var support and the active power restriction, by invoking a combination control of active and reactive powers generated by the PV inverters.

The work in [28] proposed a control strategy for the PV inverters to enhance the operational performance of the unbalanced four-wire distribution networks with high PV production, based on the optimal generation capability of reactive power and the real power restriction. An optimization problem with the objectives of voltage magnitude, network losses, PV production and active power restriction costs was addressed in previous work to find the optimum control values. In [29], a combination of droop-based inverter reactive power control technique with the active power curtailment (APC) technique was addressed to maintain the local voltage profile within pre-defined limits, in the case of large-scale PV implementation in a specific location. In this approach, the APC was employed only to improve the inductive reactive power generation at the maximum PV generation time.

Based on the literature reviews, the inverter reactive power compensation on both voltage regulation and PF improvement on building integrated PV (BIPV) systems have not been considered together. Hence, this research work is mostly focused on enabling reactive power control of a grid-connected BIPV system for automatic voltage regulation and PF improvement based on PCC load conditions. The reactive power control system is mainly subjected to the inverter capacity with the instantaneous PV power without any active power curtailment injecting to the grid. Most BIPV systems operate at unity PF where the systems inject only real power to the grid. This may have the effect of lowering the PF at the PCC of a lagging power system. Figure 1 indicates an example of this effect. The power relationship of the relevant load is given as S1 at a PF of 0.95 (18.15°) lagging. If a PV system that injects only active power at a unity PF, is installed at the PCC, the PF of the power combination S2 that is seen at the PCC will reduce to 0.75 (41.41°) lagging. Even though for a small-scale residential PV system, this cannot be neglected. Hence, it would be financially favourable to the electrical power system as well as to the consumers if the PV systems could operate with the facility of reactive power compensation at a local level.

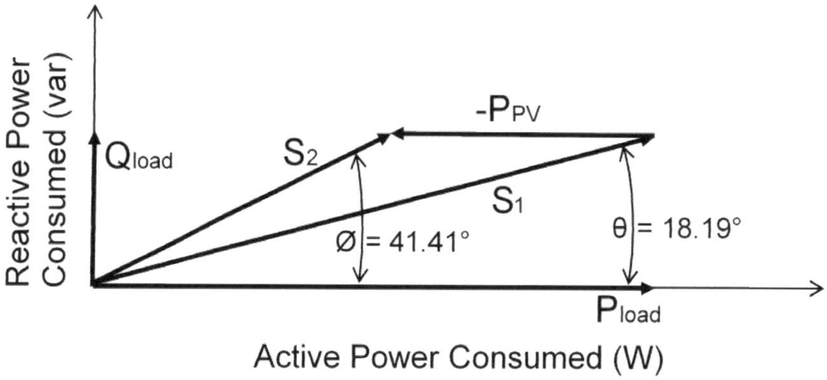

Figure 1. Active and reactive power relationship for a unity power factor PV system.

2. Modeling of PV system components

In grid-connected PV systems, the power electronic devices are used to connect the PV model into the grid because of its stochastic nature. The different PV system configurations are concerned to connect the PV modules into the grid.

Usually, the configurations can be divided into two categories, such as the PV inverters with DC/DC converter (Double-Stage) and the PV inverters without DC/DC converter (Single-Stage) [30], [31].

The double-stage PV system with a boost converter is used for this research study for boosting the PV voltage to a higher level. In this case, the maximum power point tracking (MPPT) is controlled by the boost DC-DC converter. The boost DC-DC converter is connected to the grid through a DC/AC inverter, which takes care of the grid current control and the grid synchronization [32].

Figure 2 represents the overall system configuration diagram of the grid-connected BIPV system. The system components are connected by the solid lines to represent the power flow while the control systems are connected by the dotted lines to represent the signal flow.

2.1. PV array model

The PV array block which is included in the MATLAB Simulink library is used as the PV array model. The PV array block enables to select the inbuilt preset PV model from the National Renewable Energy Laboratory (NREL) System Advisor Model. The PV array model is implemented as strings of the module connected in parallel. The string module is built up with series connected modules. As the main inputs of the PV array module, the solar intensity and the cell temperature values are considered. The SunPower PV panel is considered in the modeling and the PV model specifications are given in table 1.

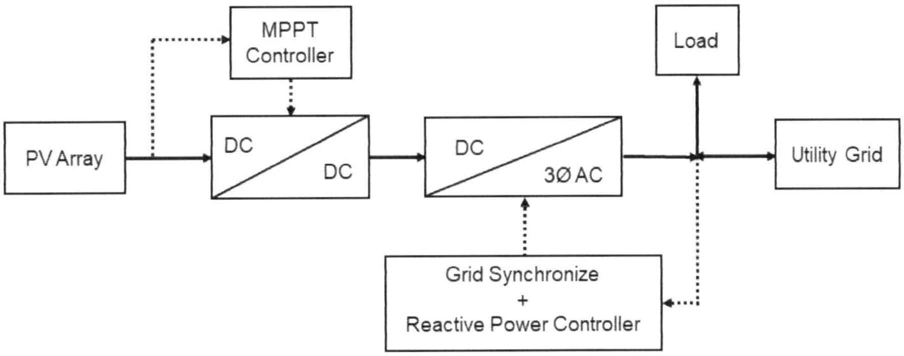

Figure 2. Overall system configuration diagram of the grid-connected BIPV system.

Table 1. PV module specifications (SUNPOWER SPR-305E-WHT-U).

PV Module	
Maximum Power (W)	305.226
Open Circuit Voltage (V)	64.2
Short Circuit Current (A)	5.96
Voltage at MPP (V)	54.7
Current at MPP (A)	5.58
Cells per module	96
PV Array Module	
Modules in series	5
Modules in parallel	5
Maximum Power (kW)	7.63

2.2. DC-DC boost converter model

To efficiently track the MPP of the PV array, the duty cycle of the converter is varied by the boost converter under all conditions. The DC bus voltage of the system is kept at 800 V. When PV array is in MPP operating condition under 1000 W/m^2 of solar irradiance and 25 °C of cell temperature, its output voltage is 273.1 V. Hence, a DC-DC boost converter is deployed to step up the PV array output voltage. The model of the boost converter is shown in figure 3. The IGBT is utilized as the power switching device in the converter.

2.3. Boost converter controller model

The gate pulse for the IGBT is generated accordingly by varying the duty ratio in order to track the MPP of the PV panel. The incremental conductance algorithm is utilized to follow the MPP voltage of the PV panel and the duty cycle is continuously changed accordingly to produce the pulses for the boost converter. The model of the boost converter controller with MPPT function is illustrated in figure 4. The initial duty cycle is set as 0.5 in the controller.

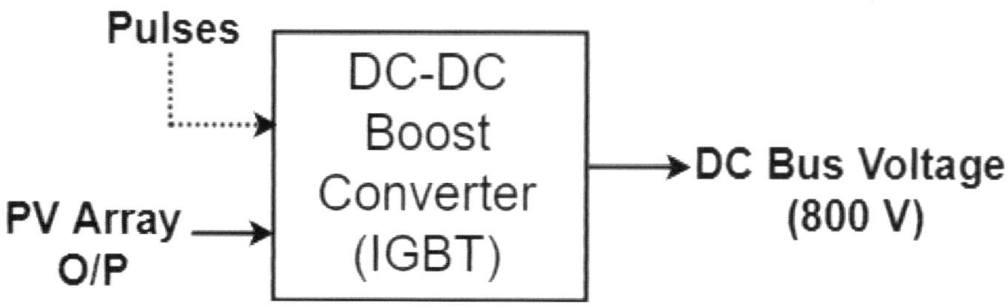

Figure 3. DC-DC boost converter model.

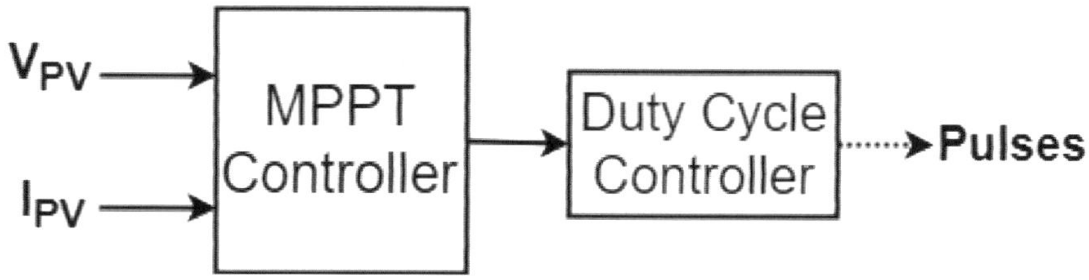

Figure 4. MPPT controller model

2.4. Voltage source inverter (VSI) model

In grid-connected PV systems, the main task of the DC-AC inverter or VSI is to convert the PV array DC power to the AC power with grid synchronization, while managing the boost converter DC output voltage in double-stage configurations. In this research study, the inverter offers the reactive power compensation service as an additional service. Hence, the two-quadrant operation of the VSI is available as shown in figure 5.

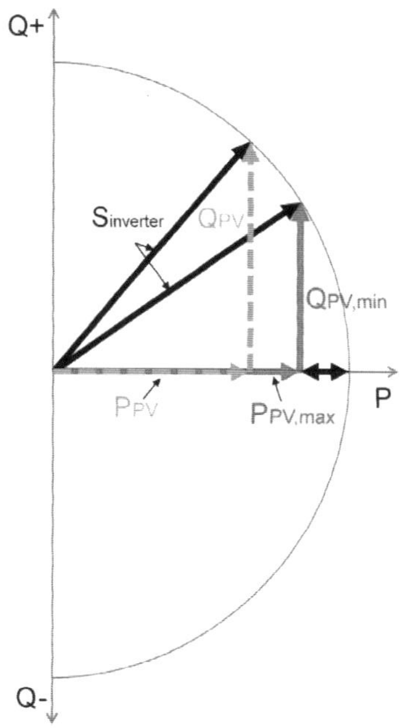

Figure 5. VSI 2-quadrant operation

The selected inverter capacity of 9.5 kVA is larger than the maximum PV capacity of 7.63 kW where the inverter maximum operating PF is 0.8 at the maximum PV capacity. Consequently, the inverter has the ability to produce reactive power at the maximum PV system operation as indicated in figure 5. The P-Q relationship of the inverter is given in equation (1).

$$Q = \begin{cases} S \times \sin\left[\cos^{-1}\left(\dfrac{S}{P_{PV,\max}}\right)\right] & ; P = P_{PV,\max} \\ \sqrt{S^2 - P^2} & ; P = P_{PV,\max} \end{cases} \tag{1}$$

Where S is the Inverter capacity, P is the instantaneous PV Power of the PV array, $P_{PV,max}$ is the maximum PV power of the PV array and Q is the reactive power of the inverter.

A three phase PWM generator is used to drive the IGBTs in the inverter. The VSI sets the input DC voltage at 800 V and transforms into a three-phase AC voltage of 400 Vrms with a frequency of 50 Hz. The harmonics caused by the inverter is filtered out by deploying an LC filter. The VSI model is depicted in figure 6.

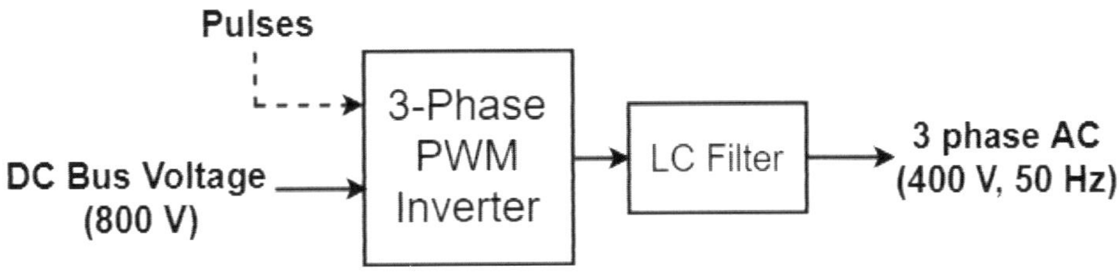

Figure 6. VSI model

2.5. *VSI controller model*

The inverter controller system shown in figure 7 is responsible for generating the PWM signals in the way to synchronize the PV system with the grid. The control system consists of a DC voltage regulator, a phase lockup loop (PLL) block, a current controller block, and a PWM signal generation block. The PLL converts the three-phase time varying system into DC system components ([V_d, V_q] are voltage components and [I_d, I_q] are current components) and measures these DC components along with the phase angle of the grid which are to be used for the grid synchronization purpose.

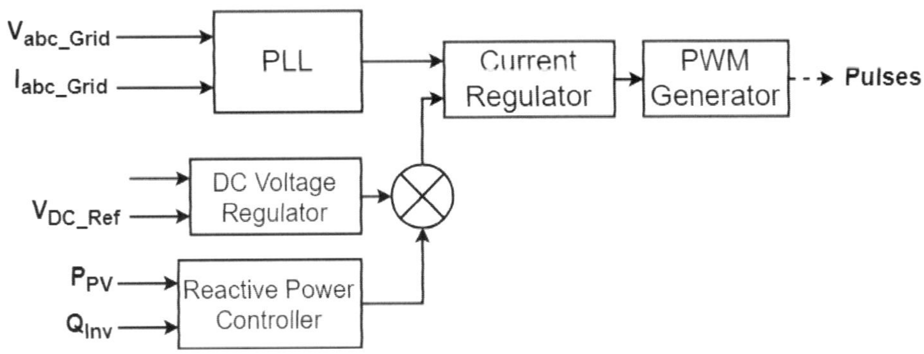

Figure 7. VSI controller model

The DC voltage regulator generates the Id current reference which is to produce the instantaneous PV power from the inverter as an active power. The I_q current reference is generated from reactive power controller to maintain the inverter output voltage within the range of ± 5% of 400 V by producing relevant reactive power from the inverter. The current controller output gives the references of V_d and V_q based on the I_d and I_q references with the measured I_d, I_q and V_d, V_q values from the PLL. The PWM signal is generated based on the three modulating signals which are generated from V_d and V_q. The reactive power controller is responsible for regulating the voltage at the PCC within the admissible limits while improving the PF. Thus, the controller guides the inverter to absorb or provide reactive power related to the required conditions. It is achieved by generating the required I_q reference to the current controller. The reactive power control is based on the voltage behavior at the PCC and subjected to the inverter limitation. The voltage at the PCC is continuously measured and determined the proper I_q reference relevant to the instantaneous PV power to manage the voltage and improve the PF at the PCC.

3. Results and Discussions

The system power flow relationship with the sign convention as represented in figure 8 is very important fact in the process of system analyzing. The general rule for the sign convention is referenced to the inverter operation. If the inverter feeds, the sign is positive and if the inverter consumes, the sign is negative. Based on this, the opposite sign is assigned for injection and consumption for both the grid and the load.

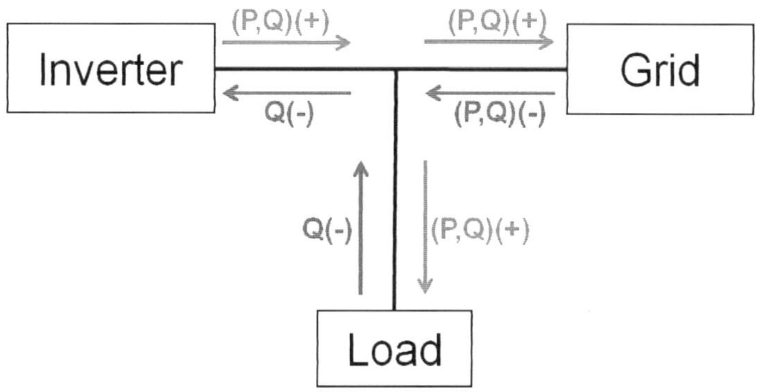

Figure 8. Relationship of the system power flow characteristic

To analyze the effectiveness of the inverter reactive power control, the implemented system is numerically investigated under real data profiles for solar and loading.

In this study, the actual solar radiation along with the cell temperature data which are measured from a Zero Energy Building (ZEB) system is utilized. The ZEB system's load profile is considered as the dynamic load condition at the PCC, where the load PF varies in the range of 0.8 - 0.95.

The system is simulated considering the solar radiation and the cell temperature profiles of the midday in the months of January and June.

3.1. System Performance Evaluation in January

The related load profile of the day is depicted in figure 9. The maximum real load of the day is more than 4 kW and the reactive power varies with respect to the load PF. Figure 10 represents the PV power characteristic of the midday in January. The peak PV power is less than 4 kW.

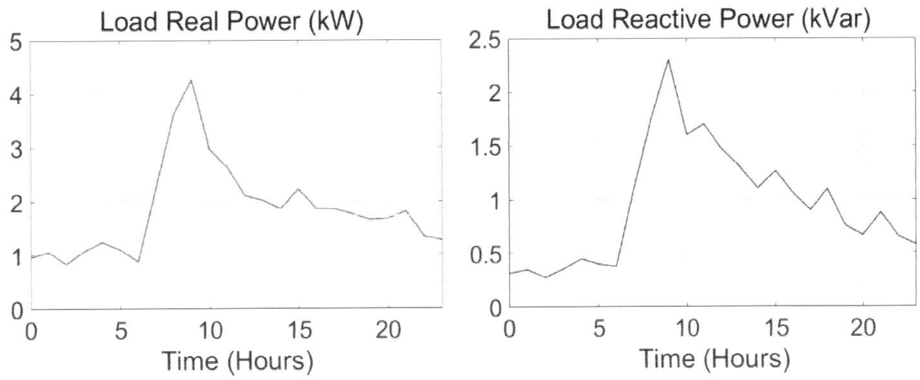

Figure 9. Load power flow characteristic

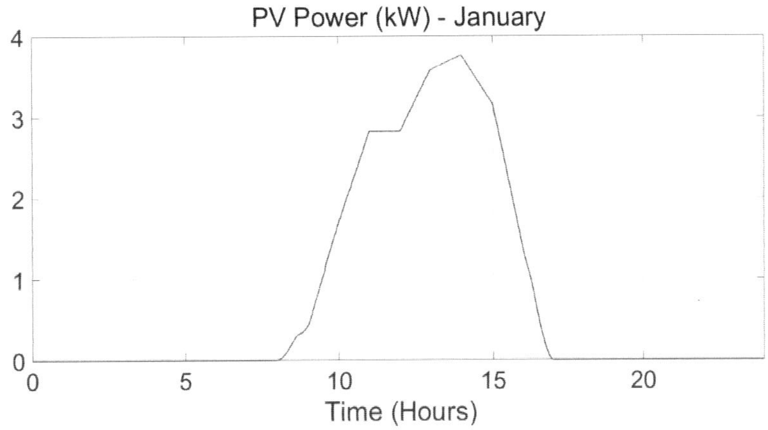

Figure 10. PV power characteristic.

The inverter power flow characteristic and the PF variation are illustrated in figure 11 and figure 12 respectively. It can be observed that the inverter var compensation is high when there is no PV power. It gradually decreases with the PV power. The inverter operating PF is in the range of 0 - 0.4.

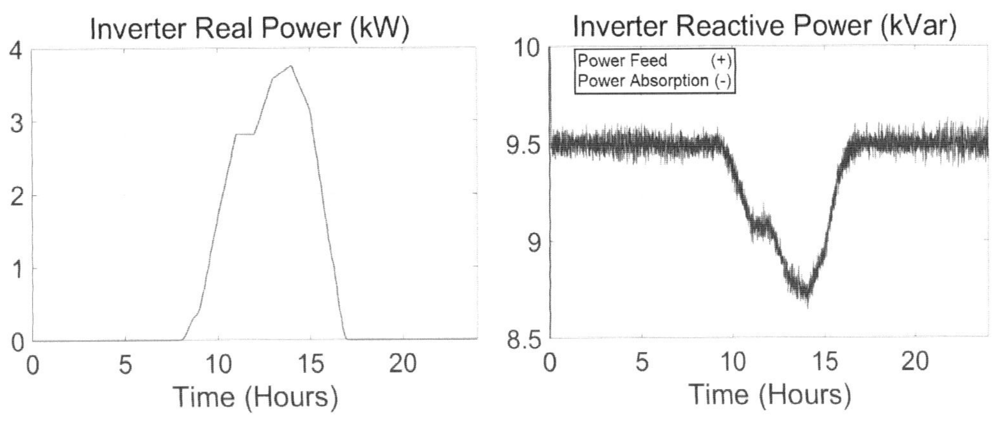

Figure 11. Inverter power flow characteristic

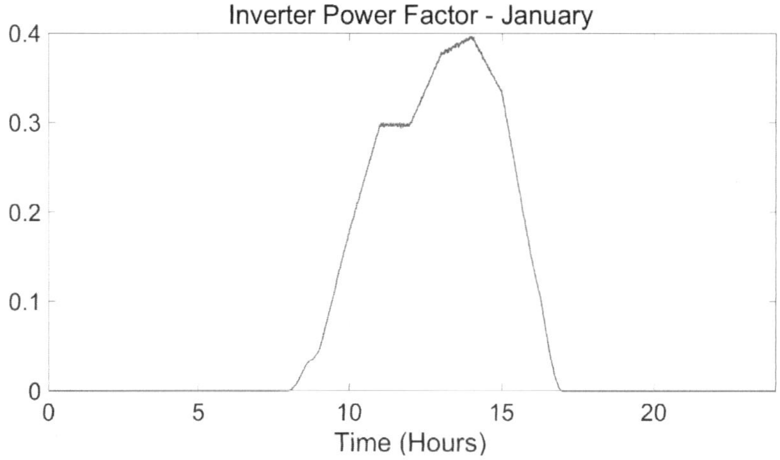

Figure 12. Inverter PF characteristic

It is observed that the grid is responsible for serving the load when there is no PV power and absorbs the excess reactive power from the inverter as can be illustrated in figure 13. The illustration of the overall system power flow characteristic is given in figure 14. The RMS voltage characteristic at the PCC is shown in figure 15.

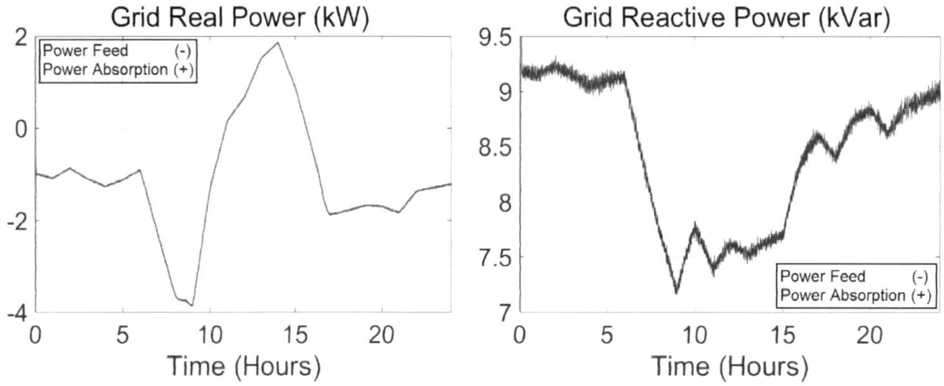

Figure 13. Grid power flow characteristic

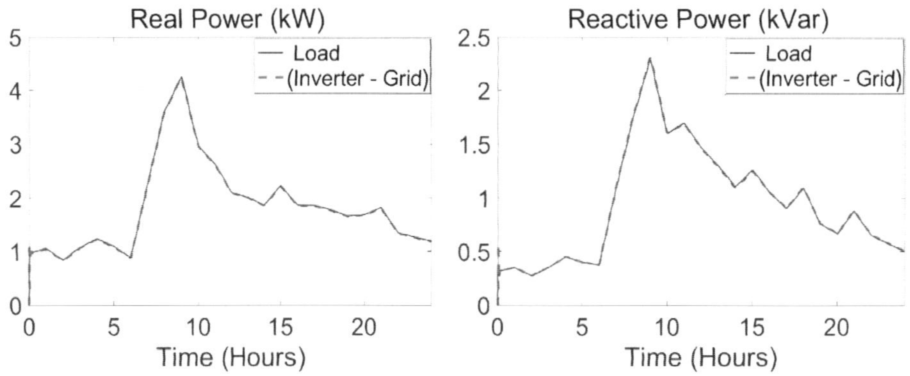

Figure 14. System power flow characteristic

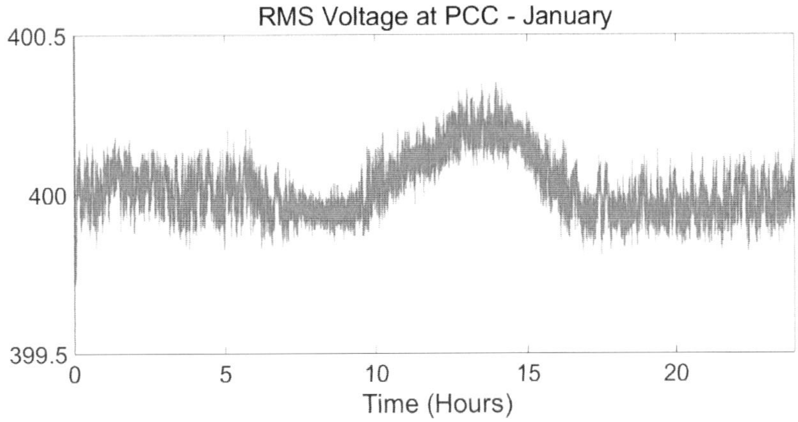

Figure 15. RMS voltage characteristic at the PCC

A slight increment is noticed in the RMS voltage when the inverter var compensation is gradually decreasing.

3.2. System Performance Evaluation in June

Generally, June is one of the months in the summer season. The system performance is explored using solar radiation and cell temperature profile of the midday in the month of June. The related load profile of the day is shown in figure 16.

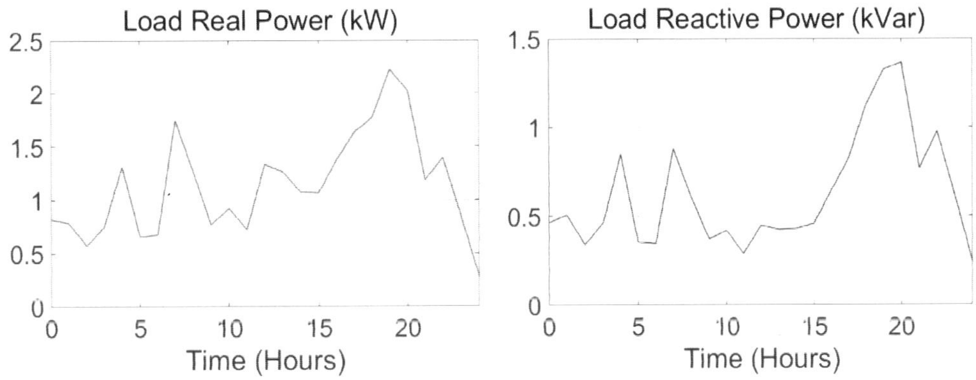

Figure 16. Load power flow characteristic

The power characteristics of the PV module and the inverter are illustrated in figure 17 and figure 18 respectively. High PV penetrations can be observed throughout the day. The inverter PF characteristic is shown in Figure 19. The inverter operates in maximum operating PF during the mid-hours while lowering the inverter var compensation.

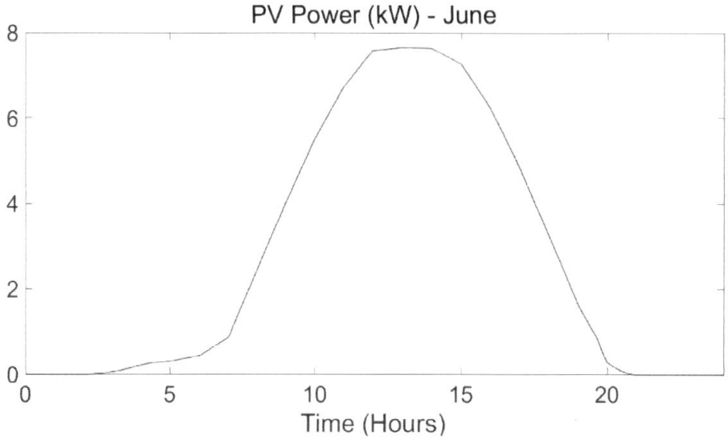

Figure 17. PV power characteristic.

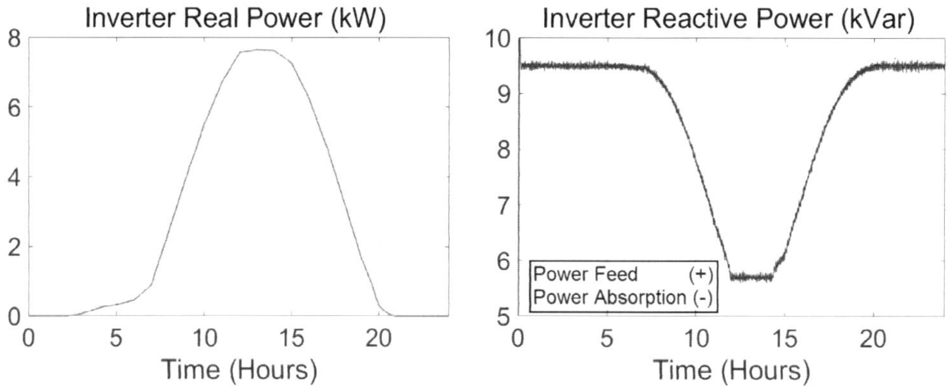

Figure 18. Inverter power flow characteristic

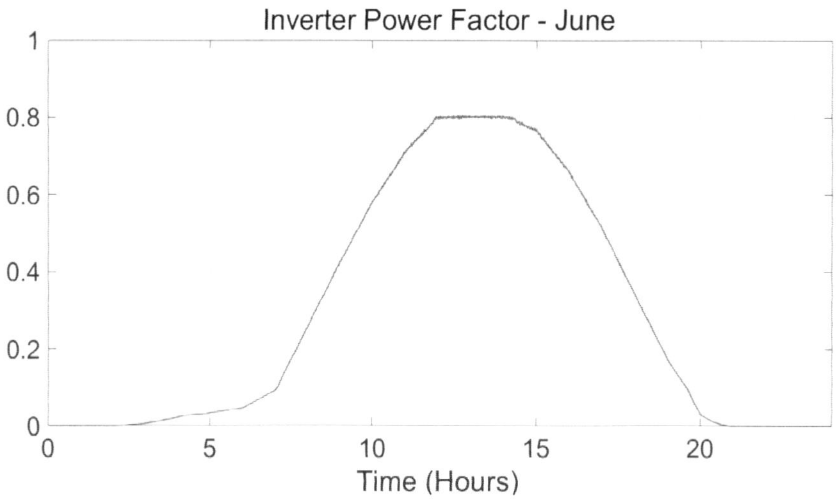

Figure 19. Inverter PF characteristic.

The grid power flow characteristic is indicated in figure 20. The overall system power flow and the RMS voltage characteristic at the PCC are represented in figure 21 and figure 22 respectively.

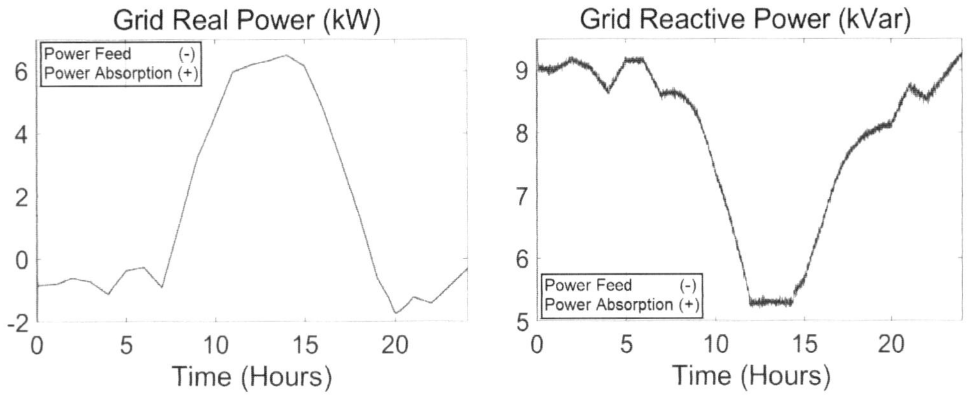

Figure 20. Grid power flow characteristic

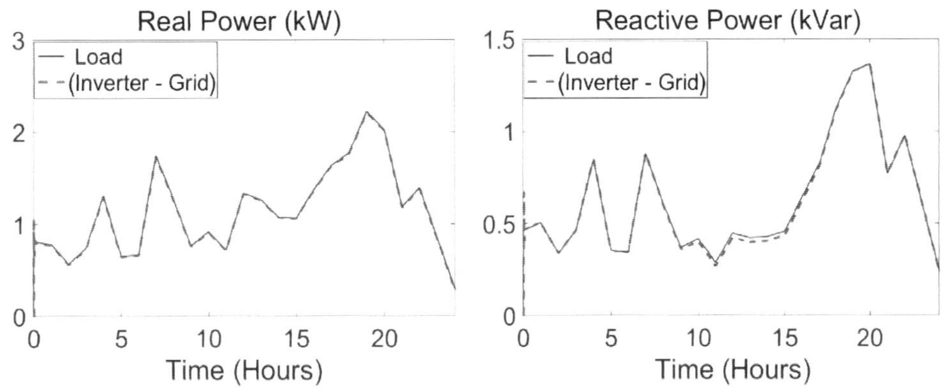

Figure 21. System power flow characteristic

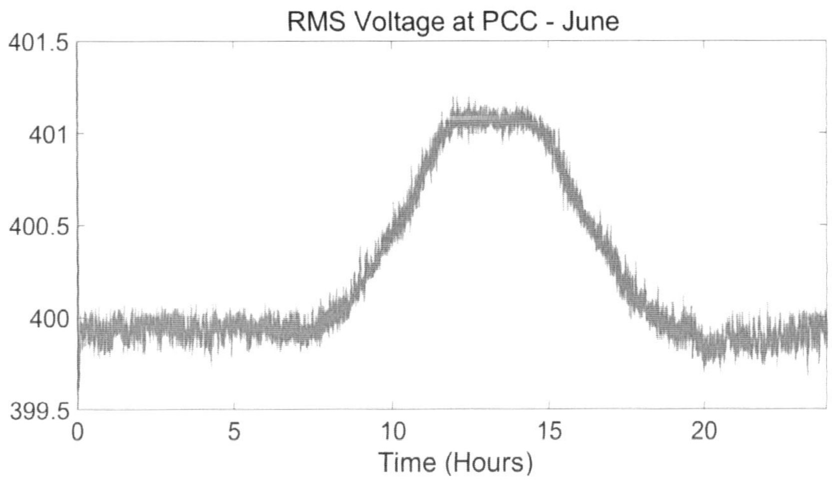

Figure 22. RMS voltage characteristic at the PCC

It can be noticed that the RMS voltage at the PCC escalates above 401 V during the mid-hours of the day when the inverter PF is high. But the overall system manages to operate subject to the limitations.

The simulation result of the overall system performance for real data indicate that the system is ready to feed maximum available power from the PV module and the reactive power subjected to the inverter capacity regardless of any loading conditions to regulate the system voltage and improve the system PF at low or high PV penetration situations where the grid fulfills the load demand when there is no PV power and absorbs excess reactive power from the inverter. This implies that the system functions with its duties where it leads to a better BIPV system operation by reducing the burden on the utility grid.

4. Conclusions

This research work presents an overall implementation of a 3-phase grid connected BIPV system with the reactive power control capability. The system model includes a PV array module with 7.63 kW capacity, a 3-phase VSI with 9.5 kVA capacity, a boost converter, a 3-phase dynamic load and a utility grid. The main aim of this work is to utilize the reactive power ability of the VSI to regulate the system voltage and improve the system PF by enabling the inverter var compensation while reducing the var compensation from the utility grid. The reactive power control is designed and implemented based on the voltage behavior at the PCC with the inverter limitations and embedded into the inverter control system. Hence, the VSI is responsible for regulating and converting the PV DC power to the AC power while producing and absorbing reactive power as an ancillary service. The boost converter ensures the operation of the PV system at MPP.

The validity and the accuracy of the implemented model were verified by comparing the system performance results with a real system. A better system accuracy was demonstrated from the results. Hence, this research study suggests that more focuses on effective utilization of the reactive power capability of a 3-phase grid-connected BIPV system capitalize on the utility grid to minimize the complexity of the control operation for mitigating the PV system effects on the system network.

References

[1] "International Energy Agency Photovoltaic Power Systems Programme", 2017.[Online]. Available: http://www.iea-pvps.org. [Accessed: 15- Apr- 2018].

[2] R. Tonkoski, L. A. C. Lopes, "Voltage regulation in radial distribution feeders with high penetration of photovoltaic", Proc. IEEE Energy 2030 Conf., pp. 1-7, 2008.

[3] Y. Ueda, T. Oozeki, K. Kurokawa, T. ltou, K. Kitamura, Y. Miyamoto, M. Yokota, H. Sugihara, S. Nishikawa, "Analytical results of output restriction due to the voltage increasing of power distribution line in grid-connected clustered PV systems", Conf. Rec. 31st IEEE Photovoltaic Specialists, pp. 1631-1634, 2005.

[4] Y. Ueda, K. Kurokawa, T. Tanabe, K. Kitamura and H. Sugihara, "Analysis Results of Output Power Loss Due to the Grid Voltage Rise in Grid-Connected Photovoltaic Power Generation Systems", IEEE Transactions on Industrial Electronics, vol. 55, no. 7, pp. 2744-2751, 2008.

[5] P. Jahangiri and D. Aliprantis, "Distributed Volt/VAr Control by PV Inverters", IEEE Transactions on Power Systems, vol. 28, no. 3, pp. 3429-3439, 2013.

[6] C. Chang, Y. Lin, Y. Chen and Y. Chang, "Simplied Reactive Power Control for Single-Phase Grid-Connected Photovoltaic Inverters", IEEE Transactions on Industrial Electronics, vol. 61, no. 5, pp. 2286-2296, 2014.

[7] W. Peng, Y. Baghzouz, S. Haddad, "Local Load Power Factor Correction by a Grid-Interactive PV Inverter" in IEEE-PowerTech, Grenoble, France:, June 2013.

[8] IEEE Std. 1547-2003, IEEE Standard for Interconnecting Distributed Resources with Electric Power Systems, IEEE, 2003.

[9] IEEE Std. 1547a-2014, IEEE Standard for Interconnecting Distributed Resources with Electric Power Systems - Amendment 1, IEEE, 2014.

[10] T. Fawzy, D. Premm, B. Bletterie, A. Gorsek, "Active contribution of PV inverters to voltage control - from a smart grid vision to fullscale implementation", Elektrotechnik und Informationstechnik, vol. 128, pp. 110-115, 2011.

[11] B. Bletterie, S. Kadam, R. Bolgaryn and A. Zegers, "Voltage Control with PV Inverters in Low Voltage Networks - In Depth Analysis of Diffeerent Concepts and Parameterization

Criteria", IEEE Transactions on Power Systems, vol. 32, no. 1, pp. 177-185, 2017.
[12] "VDE-AR-N 4105:2011-08 Forum Netztechnik/Netzbetrieb im VDE (FNN)", Power Generation Systems Connected to the Low/Voltage Dstribution Network, 2011.
[13] S. Weckx, C. Gonzalez and J. Driesen, "Combined Central and Local Active and Reactive Power Control of PV Inverters", IEEE Transactions on Sustainable Energy, vol. 5, no. 3, pp. 776-784, 2014.
[14] M. Ghasemi and M. Parniani, "Prevention of distribution network overvoltage by adaptive droop-based active and reactive power control of PV systems", Electric Power Systems Research, vol. 133, pp. 313-327, 2016.
[15] G. Nguyen and K. Uchida, "Active and Reactive Power Control Techniques Based on Feedback Linearization and Fuzzy Logic for Three-Phase Grid Connected Photovoltaic Inverters", Asian Journal of Control, vol. 17, no. 5, pp. 1522-1546, 2014.
[16] S. Ozdemir, S. Bayhan, I. Sefa, N. Altin, "Three-phase multilevel grid interactive inverter for pv systems with reactive power support capability", Smart Grid and Renewable Energy (SGRE) 2015 First Workshop on, pp. 1-6, 2015.
[17] G. Kerber, R. Witzmann, H. Sappl, "Voltage limitation by autonomous reactive power control of grid connected photovoltaic inverters", Proc. Int. Conf. Compatibility and Power Electronics, pp. 129-133, 2009.
[18] B. Bletterie, A. Gorsek, T. Fawzy, D. Premm, W. Deprez, F. Truyens, A. Woyte, B. Blazic, B. Uljanic, "Development of innovative voltage control for distribution networks with high photovoltaic penetration: Voltage control in high PV penetration networks", Prog. Photovolt. Res. Appl., vol. 20, no. 6, pp. 747-759, Sep. 2012.
[19] Y. Yang, H. Wang, F. Blaabjerg and M. Simes, "Power control exibilities for grid-connected multi-functional photovoltaic inverters", IET Renewable Power Generation, vol. 10, no. 4, pp. 504-513, 2016.
[20] G. Tsengenes and G. Adamidis, "Investigation of the behavior of a three phase grid-connected photovoltaic system to control active and reactive power", Electric Power Systems Research, vol. 81, no. 1, pp. 177-184, 2011.
[21] E. Demirok, P. Gonzlez, K. Frederiksen, D. Sera, P. Rodriguez and R. Teodorescu, "Local Reactive Power Control Methods for Overvoltage Prevention of Distributed Solar Inverters in Low-Voltage Grids", IEEE Journal of Photovoltaics, vol. 1, no. 2, pp. 174-182, 2011.
[22] R. Aghatehrani and A. Golnas, "Reactive power control of photovoltaic systems based on the voltage sensitivity analysis", in Proc. IEEE PES Gen. Meet., pp. 1-5, 2012.
[23] M. Farivar, R. Neal, C. Clarke, and S. Low, "Optimal inverter VAR control in distribution systems with high PV penetration", in Proc. IEEE PES General Meet., pp. 1-7, 2012.
[24] C. Garcia Bajo, S. Hashemi, S.B. Kjsar, Guangya Yang, J. stergaard, "Voltage unbalance mitigation in LV networks using three-phase PV systems", Industrial Technology (ICIT) 2015 IEEE International Conference on, pp. 2875-2879, 17-19 March 2015.
[25] A. Samadi, R. Eriksson, L. Soder, B. Rawn and J. Boemer, "Coordinated Active Power-Dependent Voltage Regulation in Distribution Grids With PV Systems", IEEE Transactions on Power Delivery, vol. 29, no. 3, pp. 1454-1464, 2014.
[26] K. Turitsyn, P. Sulc, S. Backhaus, M. Chertkov, "Distributed control of reactive power flow in a radial distribution circuit with high photovoltaic penetration", Proc. IEEE Power and Energy Society General Meeting, pp. 1-6, 2010-Jul.
[27] E. Dall'Anese, S. Dhople and G. Giannakis, "Optimal Dispatch of Photovoltaic Inverters in Residential Distribution Systems", IEEE Transactions on Sustainable Energy, vol. 5, no. 2, pp. 487-497, 2014.
[28] X. Su, M. Masoum and P. Wolfs, "Optimal PV Inverter Reactive Power Control and Real Power Curtailment to Improve Performance of Unbalanced Four-Wire LV Distribution Networks", IEEE Transactions on Sustainable Energy, vol. 5, no. 3, pp. 967-977, 2014.
[29] S. Ghosh, S. Rahman, M. Pipattanasomporn, "Local distribution voltage control by reactive

power injection from PV inverters enhanced with active power curtailment", 2014 IEEE PES General Meeting, pp. 1-5, 2014.
[30] F. Balaabjerg, R. Teodorescu, Z. Chen, M. Liserre, "Power converters and control of renewable energy systems", In: Proc. 6th Int. Conf. Power Electron, pp1-20. 2004.
[31] H. Haeberlin, "Evolution of inverters for grid connected PV-systems from 1989 to 2000", in Proc. 17th Eur. Photovoltaic Solar Energy Conf., Munich, Germany, Oct. 22-26, 2001, pp. 426-430.
[32] J. M. Carrasco et al., \Power-electronic systems for the grid integration of renewable energy sources: A survey", IEEE Transactions on Industrial Electronics, vol. 53, no. 4, pp. 1002-1016, Jun. 2006.

Economic Analysis of a Building Integrated Photovoltaic System Without and With Energy Storage

Pooja Sharma[1], Mohan Kolhe[2], Arvind Sharma[2]

[1] School of International Studies, Jawaharlal Nehru University, New Delhi, India

[2] Faculty of Engineering and Science, University of Agder, PO box 422, NO 4604, Kristiansand, Norway

E-mails: pooja_cos@yahoo.com, mohan.1.kolhe@uia.no, arvind.sharma@uia.no

Abstract. A grid-connected solar photovoltaic (PV) system with energy storage can help in overcoming the intermittency as well as in reducing the peak demand on the network. It also benefits in electricity bills savings. In this context, it is significant to examine the appropriate use of local energy storage (i.e. battery) connected with PV for reducing the energy supply from the grid, and also its contribution in peak demand reduction with more emphasis on reduction in annual consumer electricity bill. In this paper, a techno-economic analysis of Building Integrated Photovoltaic (BIPV) system with energy storage has been presented. A typical South Norwegian house, installed with PV, has been considered for identifying annual electricity bill savings as well as economic indicators by deploying appropriate battery capacity. It has been observed that BIPV with energy storage can be beneficial to the consumer both economically and technically, providing incentives for both consumers and investors. The time-wise limits on electricity consumption from the grid supply can also help to make the BIPV house with energy storage more economically attractive as well as near zero energy building.

Keywords – Building Integrated Photovoltaic System, Energy Storage, Energy Economics

1. Introduction

It is recommended by the European Commission that all the new buildings should have near zero-emission by 2020, and it has also been incorporated by the Norwegian system. The economical design for electricity generation using hybrid energy source PV-Biomass for an agricultural farm, and a residential community has been reported in ref [1], but it has not considered the connection of energy community with grid and adequate details on energy economics. There are many studies [2, 3] on the off-grid hybrid energy system based on solar PV, but most of them have not considered the impact of grid extension as well as potential grid tariffs in the techno-economic analysis. In most of the countries, the building integrated photovoltaic (BIPV) system has been considered, and its penetration is increasing exponentially. BIPV with energy storage can be a more appropriate option for making the buildings near to zero emission, as well as net-zero energy building from the electrical energy point of view. Also, BIPV with energy storage can contribute to demand-side management as well as with appropriate power conditioning devices can operate as a dispatchable generator within the distributed network [4]. The grid interaction performance of the BIPV system for a typical zero energy building has been reported in ref [5], but it has not considered the integration of energy storage. The reported results in [5] have been used to estimate battery energy storage capacity considering the savings in the annual electricity bill [6]. In [5, 6], the operational, economic analysis of the BIPV house with energy storage using electricity tariff has not been sufficiently addressed. In a study [7], the ratio of the installed PV peak power to the useable capacity of the battery has been used for the analysis of self-consumption, self-reliance and economic efficiency of the system. In most of the studies [2-6], the BIPV production with load profiles have not been adequality explained with techno-economic benefits to the consumer.

The appropriate energy storage (i.e. lead-acid battery) sizing for a BIPV has been proposed in the ref [6] based on electricity bill minimization using flat-rate electricity tariff. The feasibility of replacing conventional energy storage technologies with hydrogen technologies is examined for Japan [8]. Ref. [9] has examined the effect of PV- battery penetration on the cost of energy and also on the operational hours of other distributed generators but has not significantly addressed the economic benefits to the consumer. Ref [10] has studied battery capacity estimation for BIPV under different geographical conditions for the lead-acid battery using flat-rate electricity tariff. Energy storage can enhance the profitability of the grid-connected BIPV system as well as can effectively contribute to demand-side management and for improving the local energy supply for making it self-sustainable and reliable as near zero energy building. To initiate the widespread use of energy storage integrated with a PV system, various government driven support schemes have been promoted. Most of the mentioned references [1-10] have not sufficiently addressed the techno-economic analysis of the BIPV system with energy storage considering real operational PV production and the load with appropriate electricity pricing. The levelized energy cost of the PV system and the discounted payback period of PV system with market penetration are explained in ref [11] and [12] respectively.

In this work, techno-economic analysis of BIPV system with energy storage has been studied for identifying annual electricity bill savings considering electricity energy pricing. The key objective of this work is to consider the real hourly PV output with corresponding load profile and battery energy throughput for analysing the annual electricity bill of the consumer based on market electricity tariff. An operational, economic assessment of BIPV system without and with battery energy storage is going to be analysed and will be useful for not only on certain policy implications but also to highlight the appropriate incentives for investing in the energy storage.

Apart from the introduction Section 1, Section 2 of the paper presents the operational results of a typical South Norwegian House with BIPV. Section 3 compares both the technical and economic benefits of integrating the battery of energy storage, and Section 4 concludes the study by providing some recommendations and further work.

2. BIPV System Description and Operational Results

In this work, a typical South Norwegian house with BIPV is considered with an area of 154 m^2 [5]. The usable area of the roof is 57.43 $m^{2,}$ and it is facing South-West. It is considered that the installed crystalline silicon PV panels (manufactured by Sunpower SPR 230NE-BLK-I) are covering area 39.9 m^2 of the roof. A considered typical BIPV system has a capacity of 7.4 kWp with fixed roof mount with the tilt angle of 32^0 and Azimuth 48^0 [5]. The electricity production and consumption have provided significant information to assess the energy profile of a typical South Norwegian house, built on zero energy building concept [5, 6].

2.1 PV Power Production and Load Profile

The annual PV generation and load profile of a typical South Norwegian House for the year 2016 are reported in Figs.1 and two respectively. The annual grid supply concerning the load and PV production in a particular time is reported in Fig. 3. The Figs. 1 to 3 is based on the results reported in ref [5, 6].

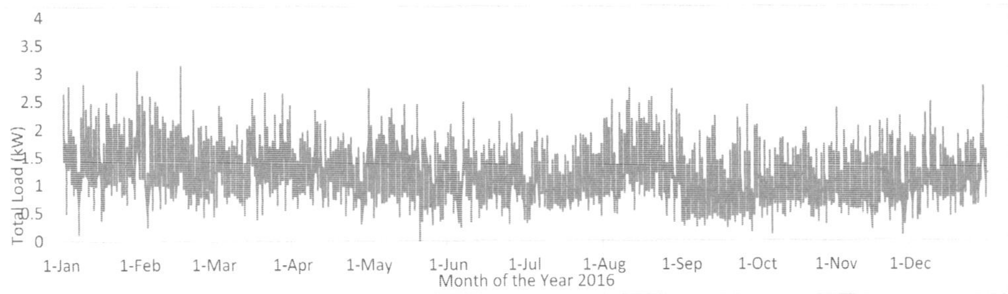

Figure 1: Annual PV production at a typical Southern Norway house

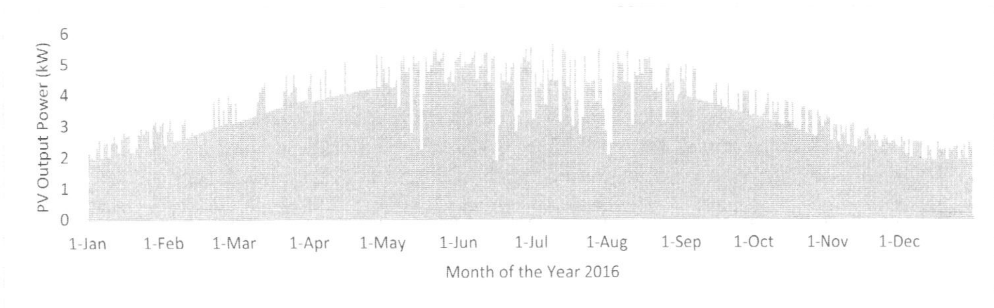

Figure 2: Annual load profile of a typical Southern Norway house

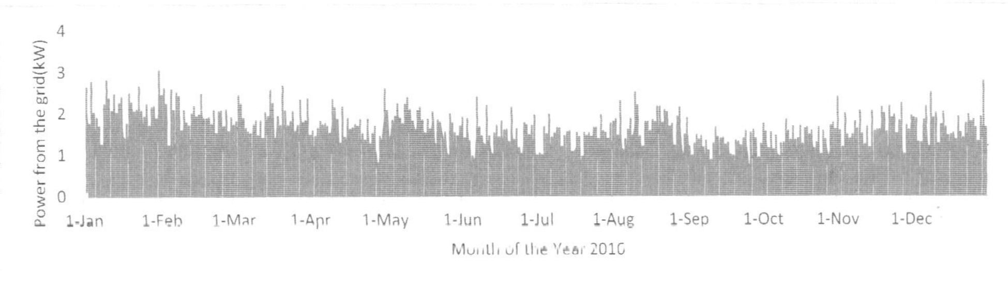

Figure 3: Annual Grid Supply to a typical Southern Norway house without energy storage

In terms of the electrical energy performance of a typical Southern Norway house, it is found that the capacity factor is 39.4%, and such BIPV system can generate out the energy of approximately 6400 kWh/yr. In the presented analysis, it is assumed that the PV system cost has been included in the housing price, and the consumer has not to pay any additional costs other than the regular mortgage of the house. Therefore, in the economic analysis, the capital cost of the BIPV system is not considered. In this work, the different purchase and sale price of electrical energy in different months of the year are considered and shown in Fig. 4. The grid purchasing price is based on ref [13], and the selling price is assumed lower than the buying price.

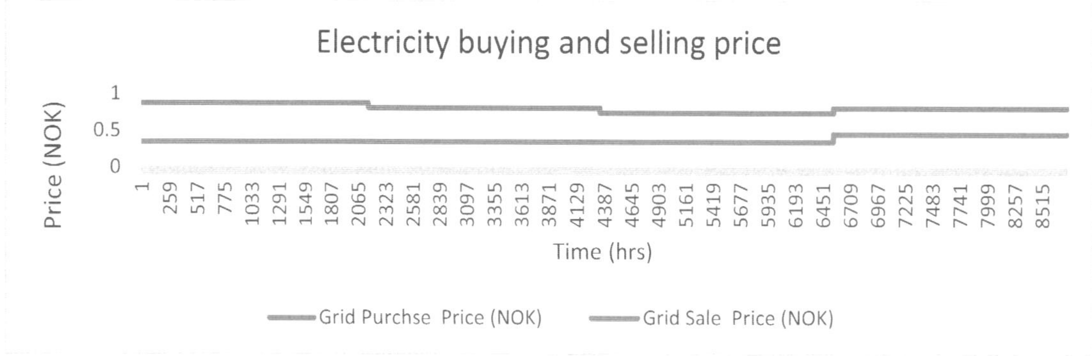

Figure 4: Grid purchase and sale price of electrical energy in the different months of a year

Through analysis, it is found that the primary load is 10308 kWh/year, with the PV production of 9860 kWh/year, and the grid electricity purchases are 6044 kWh/year, which is approximately 38%. The excess electricity of 3135 kWh/year is generated. Based on the assumptions and considered sale/purchase of electrical energy price and with consideration of remaining technical lifetime of the ten years, the net present value of BIPV system is estimated as NOK 49418, the calculated Levelized cost of energy is NOK 0.479 and the discounted payback period is 7.24 year. The levelized energy cost calculation is based on the ref [11] and the discounted payback period calculation is based on ref [12]. To reduce grid dependency and provide greater economic benefits, energy storage should be integrated with a PV system to evaluate greater economic feasibility and explore the viability of reducing grid dependency.

2.2 BIPV System with Energy Storage

The overall intermittency of the load, and also the PV production can be overcome through energy storage with BIPV system. Energy storage can also assist in demand-side management and help in strengthening the system stability. The work attempts to explore the various benefits for both economic and technical, which can be gained by integrating appropriate energy storage with BIPV. In this work, a battery energy storage (lead-acid type) of 21 kWh is considered, based on the previous work [6]. In this work, the BIPV system with lead-acid battery storage is analysed for the annual electricity bills saving, and the grid contributions are also examined. In this scenario, the annual electricity bill of the customer is reduced to 1175 NOK after integrating 21 kWh battery storage. The annual grid supply to the BIPV house with energy storage is given in Figure 5.

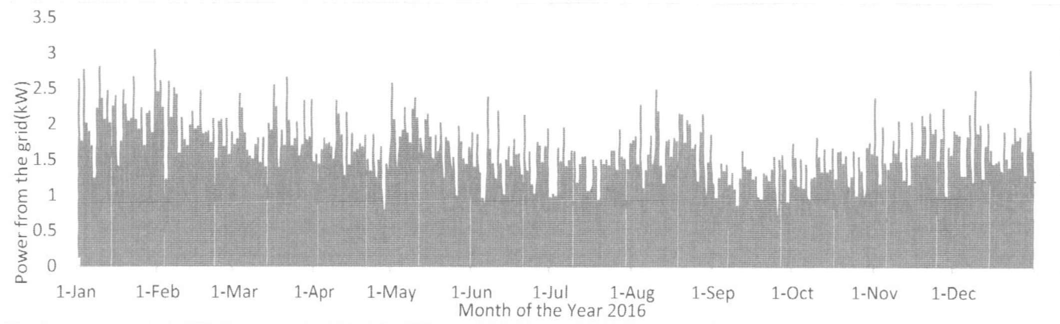

Figure 5: Annual grid supply of BIPV house with energy storage

It is evident from both Figure 3 and Figure 5, the peak load patterns (from the grid supply point of view) of the typical South Norwegian house are almost similar with and without battery energy storage. In this case, the grid purchase is 4764 kWh/year, and the battery energy throughput is 3063 kWh. The battery operating conditions have been considered with state-of-charge (SoC) from 100%

to depth-of-discharge (DoD) of 60%. The lifetime energy throughput of 1 kWh battery has been taken 840 kWh with roundtrip efficiency of 80%.

Based on the assumptions and considered sale/purchase of electrical energy price and with consideration of remaining technical lifetime of the ten years, the net present value of BIPV system with battery energy storage is estimated as NOK 38491, the calculated Levelized cost of energy is NOK 0.439 and the discounted payback period is 2.58 year. These economic indicators (Table 1) are calculated using references [11, 12].

3. Comparative Economic Evaluation of BIPV System

There are various economic indicators that can be adopted for carrying out an economic assessment of the BIPV system and battery energy storage performance [14]. In this section, the BIPV system without and with battery energy storage is evaluated economically by using certain economic assumptions. It is assumed that the BIPV system with and without energy storage has been already installed and their associated capital/installation costs are included in the mortgage of the building and not considered while computing the energy economic calculations from the consumer point of view. The considered electrical energy buying and selling prices are described in the previous section. The considered discount rate is 5.5%, and the inflation rate is 1.9% with the remaining project lifetime of 10 years. The relative comparisons of the economic indicators in both cases are given in Table 1.

Table 1. A comparison of economic indicators

Economic indicators	BIPV without energy storage	BIPV with energy storage
Net Present Cost	49418 NOK	38491 NOK
LCOE (per kWh)	0.479 NOK	0.439 NOK
Discounted payback (year)	7.24	2.69
Simple payback (year)	6.29	2.52

The annualized and operational costs (in NOK) of the BIPV system with and without energy storage are given in Fig.6.

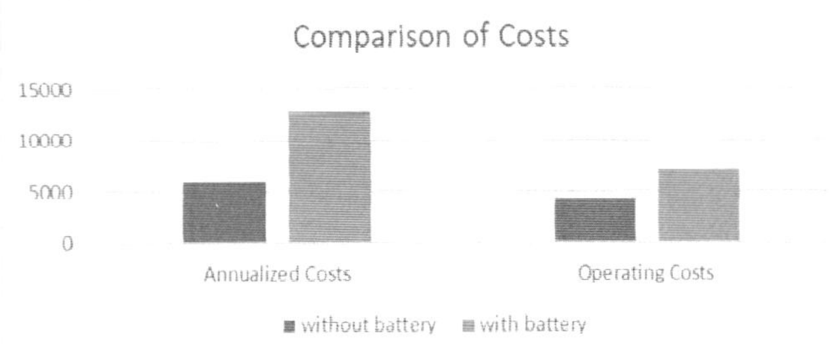

Figure 6: Comparison of capital and operating cost (in NOK) without and with energy storage

Levelized energy costs, discounted payback period etc. are good indicators of profitability of the BIPV system with energy storage, but more sensitivity analysis with economic parameters is needed concerning the battery replacements as well as operational cost analysis over the lifetime of the system. The discounted payback period is also 4.55 year less in case of BIPV system with energy storage, but more analysis is needed concerning the battery replacements as well as operational cost analysis over the lifetime of the system.

4. Conclusion

The BIPV with energy storage can play an important role not only for making a building near to net zero energy but also in demand side management considering the grid constraints as well as market electricity tariffs. In this work, a typical South Norwegian house with BIPV system is considered for potential application of energy storage. It has been observed that the BIPV system with energy storage can perform economically much better compared to its counterpart.

It is observed that a typical South Norwegian house with BIPV system with typical load profile and with considered battery capacity [6] can reduce the grid supply to 4764 kWh per year. The economic performance of BIPV with energy storage system mainly depends on the assumptions associated with PV and battery costs. In this work, it has been assumed that in both cases that the capital costs associated with PV and battery have been included in the building mortgage. Based on these assumptions, it is observed that the BIPV system with energy storage can perform economically much better compare to BIPV only system. Policies fostering investments in energy storage integrated BIPV system should be encouraged as the presented study highlights investments in BIPV system are not only beneficial for the household but also profitable for investors owing to the lower discounted payback period.

The annual electricity pricing mechanisms may change due to the load patterns as well as variation in generation scenario and also due to the introduction in the power pricing. Therefore, it is important to analyse the economic performance of the BIPV system through grid constraints and limits on-grid electrical energy purchase by the house. In further work, seasonal grid purchase limits with network constraints will be considered for doing the economic performance analysis of BIPV system with and without energy storage.

Acknowledgement:

(i) The first author is thankful to the University of Agder (Norway) for allowing doing part of her research work at Norway in cooperation with the Jawaharlal National University (India).

(ii) The University of Agder gratefully acknowledges the partial support of the Research Council of Norway and partners (Agder Energy Nett AS, Teknova As, Eltek AS) in the project "Electricity Usage in Smart Village Skarpnes" (project no. NFR-226139).

References

[1]. M.K. Shahzad, A. Zahid, T. Rashid, M.A. Rehan, M. Ali, M. Ahmad, 'Techno-economic feasibility analysis of a solar-biomass off-grid system for the electrification of remote rural areas in Pakistan using HOMER software', Renewable Energy, vol.106, pp. 264-73, 2017.
[2]. P. Mohanty, T. Muneer, M. Kolhe, 'Solar Photovoltaic System Applications: A Guidebook for Off-Grid Electrification', Springer, 2016.
[3]. M. Kolhe, K.M.I.U. Ranaweera, A.G.B.S. Gunawardana, 'Techno-economic sizing of the off-grid hybrid renewable energy system for rural electrification in Sri Lanka', Sustainable Energy Technologies and Assessments, vol. 11, pp. 53-64, 2015.
[4]. A.N. Azmi, M. Kolhe, A.G. Imenes, 'Review on photovoltaic based active generator', 9th International Symposium on Advanced Topics in Electrical Engineering (ATEE) IEEE, 2015.
[5]. A.N. Azmi, 'Grid Interaction Performance Evaluation of BIPV and Analysis with Energy Storage on Distributed Network Power Management', PhD Thesis (supervised by Prof. M. Kolhe), University of Agder (Norway), 2017.

[6]. M.A.M. Yassin, M. Kolhe, A.N. Azmi, 'Battery capacity estimation for the building-integrated photovoltaic system: Design study of a Southern Norway ZEB house', IEEE PES Innovative Smart Grid Technologies Conference Europe (ISGT-Europe), 2017.
[7]. E. Tarigan, F.D. Kartikasari, 'Techno-economic simulation of a grid-connected PV system design as specifically applied to residential in Surabaya', Energy Procedia, vol. 65, pp. 90-99, 2015.
[8]. A. Nishimura, S. Kitagawa, M. Hirota, M. Kolhe, 'Energy Assessment of Building Integrated Photovoltaics and Fuel Cell Systems: Design Study for Building(s) of Mie, Japan', Smart Grid and Renewable Energy, vol. 8, pp. 129-144, 2017.
[9]. M.A.M. Yassin, M. Kolhe, A. Sharma, S. Guard, 'Battery Capacity Estimation for Building Integrated Photovoltaic System: Design Study for Different Geographical Location(s)', Energy Procedia, vol. 142, pp. 3433-3439, 2017.
[10]. I. Koumparou, G.C. Christoforidis, V. Efthymiou, G.K. Papagiannis, G.E. Georghiou, 'Configuring residential PV net-metering policies–A focus on the Mediterranean region', Renewable Energy, vol. 113, pp. 795-812, 2017.
[11]. M. Kolhe, S. Kolhe, J.C. Joshi, 'Economic viability of the stand-alone solar photovoltaic system in comparison with the diesel-powered system for India', Energy Economics, vol. 24, pp. 155-165, 2002.
[12]. M. Kolhe, J.C. Joshi, G. Agnihotri, A.M. Shandilya, 'Market penetration and pay-back period analysis of a solar photovoltaic system under Indian conditions', vol.1, 2002.
[13]. Statistics Norway, Electricity prices at the end-user markets quarterly, 2016.
[14]. N. DiOrio, A. Dobos, S. Janzou, 'Economic Analysis Case Studies of Battery Energy Storage with SAM', National Renewable Energy Laboratory, NREL/TP-6A20-64987, 2015.

IOP Conf. Series: Materials Science and Engineering **605** (2019) 012014 doi:10.1088/1757-899X/605/1/012014

Precipitation of γ' Phase of DD6 at Unequal Service Temperatures

Yao Huang[a,b], Ning Li[a,b], Hui Zhang[a,b], Renheng Han[a,b], Chengzhi Zhao[a,b,*], and Hexin Zhang[a,b,*]

[a] College of Materials Science and Chemical Engineering, Harbin Engineering University, Harbin 150001, China
[b] Key Laboratory of Superlight Materials and Surface Technology, Ministry of Education

E-mail: zhaochengzhi@hrbeu.edu.cn, zhanghx@hrbeu.edu.cn

Abstract. In this paper, the DD6 alloys after complete heat treatment were heat-treated at 900, 1000, 1100, 1200, 1300 °C for 1 h, then was air cooled. Microstructures treated by different temperatures were investigationed by OM and FESEM. The microstructures evolution of alloys were estimat using JMatPro software and Image-Pro Plus. Conclusion illustrates the size of γ' phase increases gradually after the DD6 alloy heat-treated at 900, 1000, 1100 and 1200 °C for 1 h, and the size reaches the maximum after heat-treated at 1200 °C. Most of the γ' phase dissolves and then precipitate the irregular small γ' phase, about 0.20 nm after preservation at 1300 °C. The size distribution of the γ' phase turned into more and more uniform, and the best distribution state is achieved after preservation at 1100 °C; The matrix channel of the microstructure of DD6 widens gradually with temperatures increasing and reaches a maximum value at 1100 °C, about 65 nm. The matrix channel is narrowed gradually, and the most part of γ' phase has serrated γ/γ ' phase surface after heat treatment at 1200 °C. After heat preservation at 900 °C, volume fraction of γ' increases contrast that after complete heat-treatment; after heat-treatment at temperature degree of 1000-1200 °C, γ' volume fraction decreases gradually; γ' phase volume fraction increases again at 1300 °C. The different microstructure characteristics of DD6 after heat-treatment at unequal temperatures are obtained, indicating that the using temperature has a significant effect on the microstructure of DD6 alloy.

Keywords: DD6 alloy, single crystal superalloy, γ ' phase, microstructure evolution.

1. Introduction

Nickel-based single crystal superalloys are extensively expended on the manufacturing of turbine blades of gas turbines and jet engines, owing to their superior megathermal resistance, oxidation resistance saving good erosion rejection [1-3]. The future casting process of engine blades directly determines the performance of the engine and it is also a prominent symbol of the level of the national aviation industry. Therefore, further improving the service life of engine turbine blades under high degree of heat and high voltage have become the top priority of the current problem [4]. DD6 are independently studied by China. It has the advantages of high temperature strength, good comprehensive performance, stable structure and good casting process performance. Its tensile, long-lasting, anti-oxidation and heat-resistant corrosion properties have reached or exceeded the

performance of second-generation single-crystal superalloys PWA1484, René N5, CMSX-4, which are widely used abroad [5-6].
The γ' phase plays a major role in intensify the superalloys and it is completely coherent with the matrix γ phase, as indicated in figure1. The size, shape, quantity and distribution of γ' phase has a great influence on the performance of DD6 superalloys [7-8]. The single crystal alloy is produced by directional solidification and spiral grain selection methods, and the solute is redistributed during the preparation process because crystallization rate is slow. The dendrites of the as-cast alloy are enriched with a large number of Al and Ta. They belong to γ' forming elements. The γ forming elements for instance W and Re are concentrated in the dendrites, which makes the single crystal superalloy during solidification. The dendritic morphology is intact and there is obvious dendrite segregation, which significantly effects the performance of DD6. Therefore, heat preservation of alloy is necessary [9]. The heat treatment can not only reduce the dendrite segregation, eliminating the eutectic structure in the alloy, but also dissolve then re-precipitate γ' to obtain the finer, uniform and regular phase, and improve the performance of DD6. The heat treatment process of the single crystal superalloy is matched with the alloy composition design and other processes to make the single crystal alloy reach the optimum state. The heat treatment of single crystal superalloys is usually multi-stage heat treatment so as to eliminate segregation and eutectic structure towards the utmost extent, improving the initial melting temperature of the alloy, finally achieve a best collocation of the number, shape and composition of γ' phase. Thus, it is indispensable to conduct examination on the affect of high operating temperature on microstructure of superalloys. Standard heat-treated DD6 alloy was heat-treated at distinct temperatures to simulate alloy's service temperature. The microstructure evolution characteristics of the alloy at different temperatures were analyzed. The development mechanism of alloy structures were discussed by phase diagram calculation.

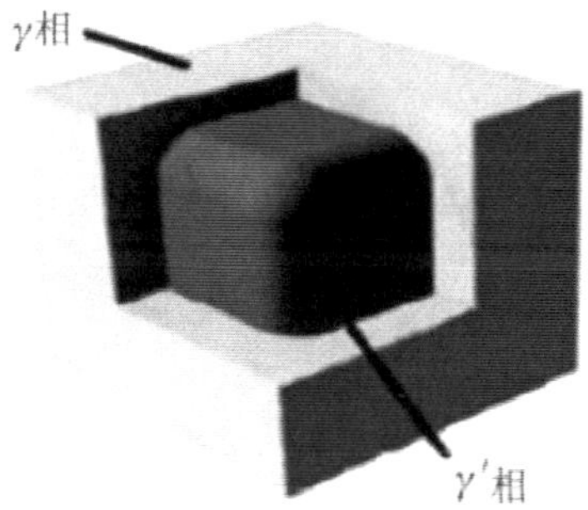

Figure 1. Two phase structure of DD6 alloy

2. Experimental procedures

Nominal composition of Nickel-based single crystal superalloy was employed as shown in Table 1. [001] oriented DD6 single crystal superalloy test bars were prepared by spiralgrain selection methods and directional solidification. The preferred <001> direction was within 10° from the direction of the principal stress axis for investigation from the as-cast specimen bar. The specimens was completely heat treated first and specimens were kept at 900, 1000, 1100, 1200, and 1300°C for 1 h individually, and then air-cooled specifically. Figure 2 demonstrates the heat-treated system. Heat treatment structures and microstructures of the alloy treated with different temperatures were inspect by OM and FESEM. The volume fractional number of γ' was survey and evaluate by Image-Pro Plus. The thermodynamic calculation of the alloy was carried out by JMatPro software.

Table 1. Chemical compositions of DD6 alloy (mass fraction/%)

Co	W	Ta	Al	Cr	Mo	Re	Nb	Hf	C	Ni
9	8	7.5	5.6	4.3	2	2	0.5	0.1	0.006	Bal.

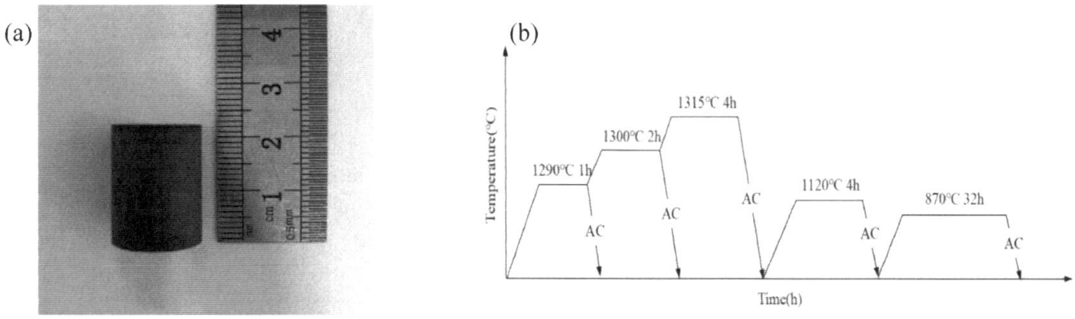

Figure 2. DD6 alloy specimen and schematic illustration of standard heat treatment schedules.

3. Ruselts and analysis

3.1. Standard heat-treated organization of alloy

Figure 3 shows the standard heat treatment conformation of DD6 after solution heat treatments and subsequent aging heat treatment. Since the micro structure of the as-cast superalloy contains more γ and γ' eutectic, dendritic segregation and rough γ' phase, these structures influence mechanical performance of alloys at high temperature, therefore the alloys must be solution heat treatments to dissove γ/γ' eutectic and minimize chemical heterogeneities. The aging treatment is then carried out to adjust the γ' phase structure to a suitable shape, quantity and size. It can be seen from the macroscopic structure shown in figure 3a that the morphology of the as-cast dendrite is not obvious. The dendritic and interdendritic are homogenized while the γ/γ' eutectic is also eliminated substantially when completely heat treated. Presented in table 2, the γ' phase is well-formed to a degree of 0.44 μm, the volume fraction of which is approximately 61%, and the average matrix channel width of the γ phase is 57 nm after standard heat treatment. The calculation results are the same as those of Shi Zhenxue et al. [10].

Figue 3. Macro and micro structure of DD6 after standard heat treatment.

3.2. Microstructures of DD6 after heat treatmenrets at different temperaretures

Figure 4 shows that γ' phase structure of DD6 at distinct temperatures for 1h. Figure 4 illustrates that after alloy is kept at 900 °C for 1 h, the γ' phase still maintains cubic form, and the degree of cubicization and size are basically unchanged; the matrix channel is more straight than before, and volume fractional number of γ' also show no significant change contrast the complete heat treatment. After 1000 °C heat treatment, the γ' phase in DD6 superalloy still maintains cubic shape, and the degree of cubicization increases slightly. The size of the γ' phase begins to increase slightly about 0.52 nm; the matrix channel becomes straight and significantly wider than complete heat treatment. The γ' phase is still cubic, but the phase becomes larger, about 0.57 nm and its size and homogenization reach the best state after heat-treated at 1100 °C; the matrix channel is more straight, the volume fractional number of γ' further grows. The size of γ' continues to increase to 0.6 nm and the degree of cubication increases but the degree of homogenization decreases; the interface of γ'/γ phase is serrated after heat preservation at 1200 °C for 1 h. The most part of γ' phase incomplete solution then reprecipitate into irregular small γ' phase after heat-treated at 1300 °C.

3.3. Quantitative analysis of different organization of DD6

The quantitative analysis of the fine structure of the DD6 at distinct degree of heat treatment was carried out and compared with the standard heat-treated microstructure of figure 3, the results are shown in Table 2. The size of γ' phase grows with increasing temperature, the volume fractional number of γ' decreases gradually within the range of 900 °C to 1200 °C. The magnitude and volume percentage of γ' improve when the temperature is 1200 °C-1300 °C. The volume fractional number of γ' phase grows to distinct extents after heat treatments at distinct temperatures. The morphology of γ' was not effected much after being kept at 900-1100 °C for 1h respectively, but it changed greatly after being kept at 1200 and 1300 °C for 1h.

Table 2. Microstructure characteristics of DD6 alloy after different temperature treatment

	Primary γ' phase size/μm	Volume fraction of Primary γ' phase/%	Morphology of primary γ' phase	Space of γ matrix channel/nm
Standard heat-treatment	0.44	61	Cubic	57
900	0.45	65	Cubic	54
1000	0.52	64	Cubic	59
1100	0.57	62	Cubic	65
1200	0.62	55	Cubic, most with serrated γ/γ' interface	56
1300	0.20	59	Irregular cubic	31

4. Discussion

4.1 Thermodynamic calculation and phase equilibria

The differences in microstructure between DD6 alloys at different temperatures are due to the fact that the alloys are heat treated at different temperatures, resulting in differences in the growth, dissolution and form sediment of γ' phase. Calculation software JMatPro and its corresponding nickel-based alloy database are used to calculate the equilibrium phase of the DD6 alloy, presented in figure 5. As can be seen from the figure, in the temperature extent of 900 to 1300 °C, DD6 alloy is mainly composed of γ matrix, γ' phase and a bit of carbide. The γ' content in the equilibrium state of the alloy is substantially zero at 1315 °C; and the γ' content in the equilibrium state in DD6 is 72% below 600 °C. Therefore, when the DD6 single crystal superalloy is cooled to room temperature after solution treatment at 1315 °C, the γ' phase forming element in the γ matrix is highly supersaturated, which causes a large amount of γ' phase precipitation during cooling. The content of C is extremely small in the alloy, so the content of C changes slightly with temperature increasing. Figure 5 shows that the content of the γ' phase is re-dissolved to the γ matrix gradually as the temperature increasing, and the complete re-dissolving temperature of the γ' phase is about 1300 °C.

4.2. microstructure evolution of DD6 alloy

The DD6 alloy is heat-treated at 900-1200 °C respectively. Since heat treatment temperature below the large re-dissolving temperature of γ' phase of 1270.1 °C[10], there are two processes are carried out simultaneously. One is the growth of γ' phase, the other is the re-dissolution of γ' phase. γ' phase will grow according to the Ostward method, the large γ' phase relatively will grow, and the small γ' phase will dissolve. γ' phase growing up follows the LSW roughening theory [11]:

$$\sqrt[3]{(r_t^3 - r_0^3)} = K\sqrt[3]{t} \quad (1)$$

where r_t is the average radius of the precipitated particles after aging, r_0 is the average radius of the precipitated particles before aging, k is the coefficint related to the aging temperature, and t is aging time.

According to the Arrhenius law [12]:

$$D = D_0 exp[- Q/(RT)] \quad (2)$$

where D is diffusion coefficient, D_0 is solute diffusion constant, R is molar gas constant, Q is diffusion activation energy and T is thermodynamic temperature.

Equation (2) illustrates temperature is the most important factor effecting the diffusion of elements in the alloy. The higher the temperature, larger the diffusion coefficient is, the diffusion rate of the elements is faster, γ' phase grows faster. Therefore, in the case where the heat treatment time is the same, the higher the temperature of the heat treatment, the size of γ' is larger. While growth rate of γ' phase is significantly greater than dissolving ratio of γ' phase . It can also be seen from Table 2 that when the temperature is 900 -1100 °C, the size of the γ' phase increases gradually. Part of the γ' phase begins to dissolve, the width of matrix channel increases during heat preservation process. When the temperature is raised from 900-1100 °C, the number of γ' re-dissolved grows, and the matrix channel becomes wider. When the heat treatment temperature achieves 1200 °C, which near the large amount of re-dissolving temperature of the γ' phase. Therefore, when DD6 is kept at 1200 °C for 1 h, the amount of re-dissolving is also large, and most of the γ' phase begins to dissolve. During the cooling process, the precipitation way of the γ' is different from which is at the lower temperature. γ' phase at 1200 °C is dependent on the γ' phase which has been precipitated by aging. Since the growth mode has a heterogeneous nucleation, this kind of growth is easier to carry out. MIYAZAKI et al. [13] pointed out the γ' phase has high elastic strain energy locally, re-precipitated γ' will precipitate along the (011) plane to reduce the energy that considering the interfacial energy and elastic strain energy of γ' phase anisotropy. Therefore, the precipitation is directionality, and the precipitation along the (011) plane grows faster, which causes the γ/γ' phase interface to appear serrrated, as shown in figure4 d2 and d3. 1300 °C is higher than the complete re-dissolving temperature of the γ' phase and less than the junior melting temperature of DD6, the γ' phase will dissolve into γ matrix completely at 1300 °C for 1 h. Fine and irregular γ' phase is then re-precipitated during the cooling process.

Figure 5 demonstrates that the content of the γ' phase in equilibrium state of the alloy is almost zero above 1300 °C; when it is below 600 °C, γ' content is about 70%. After the alloy was preserved at 1315 °C for two hours then air cooled, the γ' forming elements are supersaturated, causing γ' precipitation in a large amount during cooling. The content of γ' reached 61% after secondary aging. The volume fractional number of γ' increases at the beginning then decreases final increases as temperature increasing. The magnitude and shape of new γ' phase with original γ' phase growing, dissolved and then re-precipitated grows are different when the alloy is preserved at distinct temperatures. Therefore, volume fractional diversification of γ' phase needs further research. Large re-dissolving temperature of γ' phase is 1270.1 °C. Growth of γ' phase is main factor at 900 °C, the matrix channel becomes smaller, so the volume fractional number of γ' increases. While temperature is carried out at 1000 °C and 1100 °C which near the large amount of γ' phase, dissolution of γ' becomes the major factor, and the matrix channel becomes larger. Therefore, volume fractional number of γ' phase is reduced. When DD6 preserves at 1200 °C, the heat treatment temperature is higher, and the γ' phase re-dissolved amount continues to increase, but new γ' phase re-precipitates by original γ' phase during subsequent air cooling. In this occasion, volume fractional number of the γ' phase increases gradually. At 1300 °C, γ' phase is completely dissolved, and γ' phase which re-precipitates during subsequent air cooling is also relatively increased, so volume fractional number of γ' phase is increased.

5. Conclusions

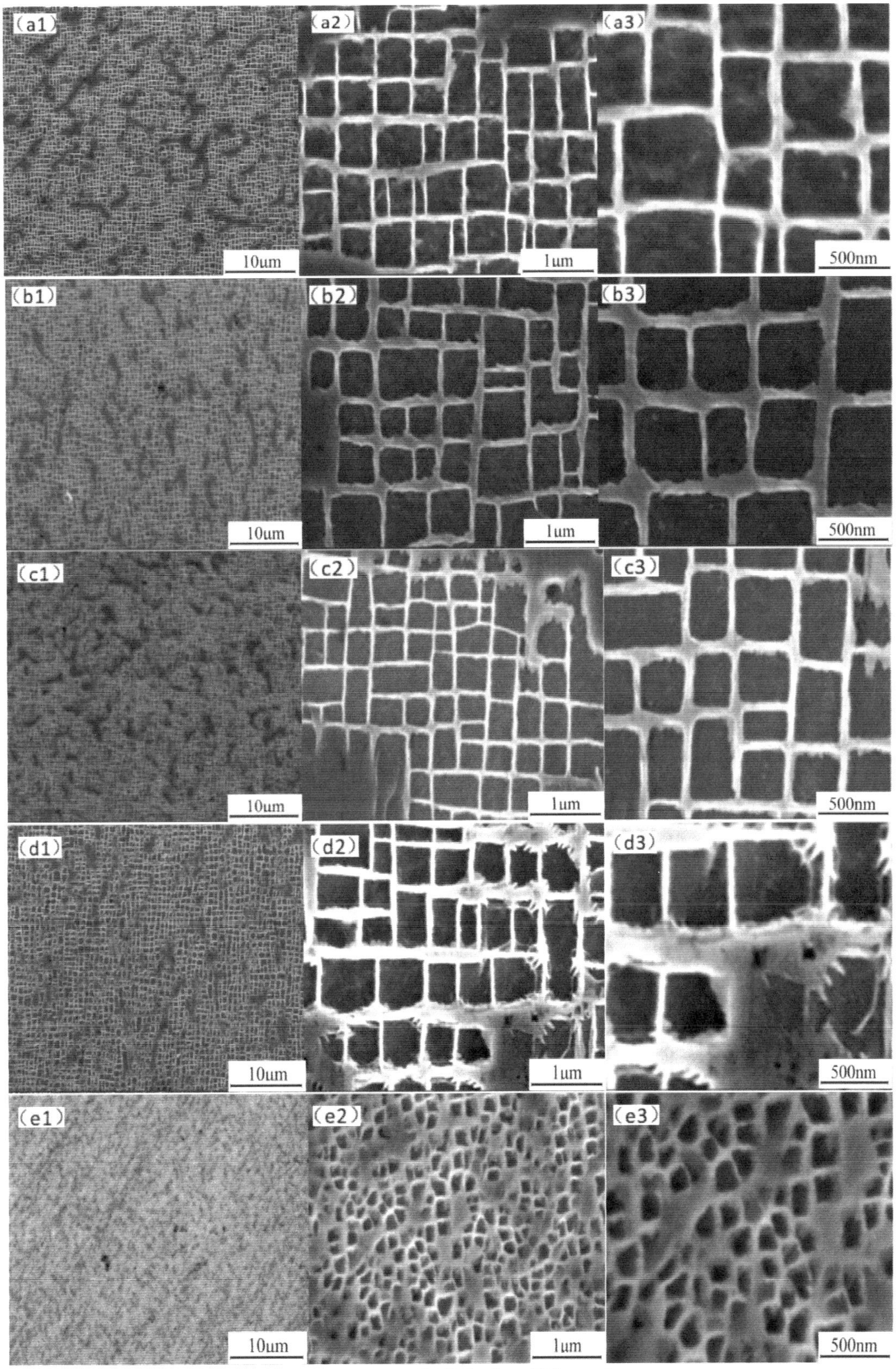

Figure 4. Micro structure of DD6 after different heat-treatment.

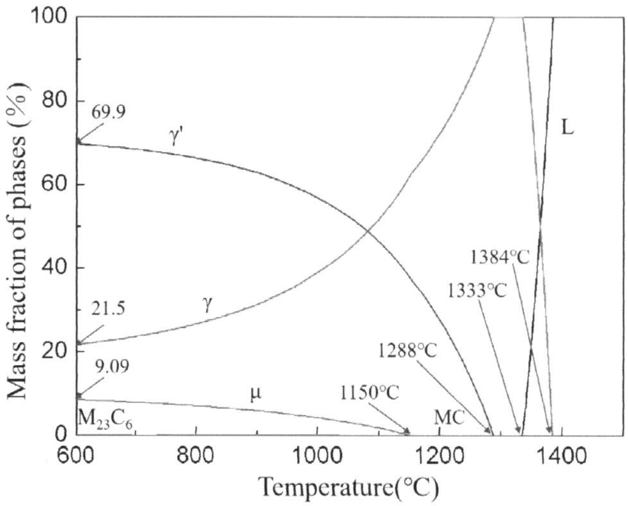

Figure 5. Relationship between equilibrium phase and temperature of DD6 alloy.

- After standard heat treatment, the γ' phase is well formed. The size of the γ' gradually increases at 900-1100 °C. when alloy preserves at 1200 °C, the size of the γ' increases significantly, the most of γ/γ' surface is serrated. The γ' phase is completely dissolved, and then in to a fine, irregular γ' phase in the process of subsequent cooling after heat-treated at 1300 °C.
- The volume fraction of γ' phase first increases and decreases and then increases, during the heat treatment from 900 °C to 1300 °C. When heat-treated at 900 °C, the growth of the γ' phase is the main factor, the matrix channel is narrow, and volume of the γ' phase grows. When the degree of heat maintains at 1000-1100 °C, the amount of γ' phase re-dissolved increases gradually and the matrix channel becomes wider and volume fractional number of γ' phase declines. γ' phase is further dissolved back after heat-treatment at 1200 °C, however during cooling, new γ' is re-precipitated by the original γ' phase, the matrix becomes wider, and volume fractional number of γ' phase decreases again. The γ' dissolves completely, a new fine and irregular γ' phase re-precipitates during cooling process after heat-treatment at 1300 °C. The matrix channel is narrowed and the γ' phase volume fraction increases.
- The micro structure of DD6 heat treated at distinct temperatures is significantly different, indicating that the operating temperature has a great affect on the microstructure of the alloy.

Acknowledgements

This paper is funded by the International Exchange Program of Harbin Engineering University for Innovation-oriented Talents Cultivation.

References

[1] Shi Z X, Liu S Z and Wang X G 2014 *T. Nonferr. Metal. Soc.* 21 2536
[2] Pillai R, Ackermann H, Hattendorf H and Richter, S 2013 Corros. Sci. 75 28
[3] Tian S G, Li Q Y and Guo Z G 2014 *T. Nonferr. Metal. Soc.* 24 668
[4] The editional committee of China aefonautical materials handbook 2002 *China aefonautical materials handbook 2nd ed* (Beijing: China Standard Press) pp 812-813
[5] Tang D Z, J R Li, Z T Wu, Liu S Z and Wu X R 1999 *J. Mater. Eng.* (3) 8
[6] Li J R, Jin H P and Liu S Z 2007 *J. Mater. Eng.* 25 22
[7] Li J R, Xiong J C and Tang D Z 2012 *Advanced High Temperature Structural Material and Technology* (National Defense Industry Press) pp 144-153
[8] Wang J, Zhang L T, Chen K, Nairong S and Aidang S 2011 *T. Nonferr. Metal. Soc.* 21 1513
[9] Balikic E, Raman A and Mirshams R A 1997 *Metall. Mater. Trans. A* 28 1993

[10] Shi Z X, Liu S Z, Xiong J C and Li J R 2015 Microstructure evolution behavior of DD6 single crystal superalloy at different using temperatures [J]. *T. Nonferr. Metal. Soc.* (11) 3077
[11] Lifshitz L 1961 The kinetics of precipitation from supersaturated solid solutions [J]. J. Phys. Chem. Solids 19 35
[12] Feng D 1998 *Metal physics* (Beijing: Science and Technology Press) pp 492–493
[13] Miyazaki, Toru, Koyama and Toshiyuki 1994 *Acta Metall. Mater.* 42 3417

The Influence of Mg/Cu on Tensile Properties of ZA27 Alloy

Hui Zhang [a,b], Yanqing Zhu [a,b], Ning Li [a,b], Yao Huang [a,b], Chengzhi Zhao [a,b,*], and Hexin Zhang [a,b,*]

[a] College of Materials Science and Chemical Engineering, Harbin Engineering University, Harbin 150001, China
[b] Key Laboratory of Superlight Materials and Surface Technology, Ministry of Education

E-mail: zhaochengzhi@hrbeu.edu.cn, zhanghx@hrbeu.edu.cn

Abstract. In this paper, the addition of magnesium and copper was changed to investigate the effect of magnesium and copper on the tensile properties of ZA27(Zn-27Al) alloy. The experimental results showed that Zn-27Al-0.8Mg has the largest tensile strength of 294.61 MPa. The fracture of Zn-27Al-xMg alloy was mainly related to the newly precipitated magnesium phases (β+β′). For the Zn-27Al-xCu alloy, when the amount of copper added was 0.4wt.%, the tensile strength reached 292.58 MPa, and when the amount of copper added was 2.0wt.%, the elongation even reached 80%, indicating that the alloy has good plasticity. The fracture of the Zn-27Al-xCu alloy was mainly related to defects in the structure and the second phase particles.

Keywords: Zinc-aluminum alloy; Tensile properties; Elongation; Second phases

1. Introduction

As a new light material, zinc-aluminum alloy had many excellent properties, such as higher hardness, better wear resistance and good damping performance. The material was used in the industrial field, which can effectively solve the problems of noise and vibration, and it can also be beneficial to achieve the goal of lightweight vehicles to a certain extent. So it has great prospects in industrial applications such as ships, aerospace, and automobiles.

In the past, people were mainly devoted to studying the properties of zinc-aluminum binary alloys[1～5] and the effect of rare earth elements on the properties of zinc-aluminum alloys. Ren et al. proved that Y increased the yield strength and hardness of Mg-Al-Zn alloys[6][7]. Jia Wenxu et al. studied the characteristics of Gd on Cu-Zn-Al shape memory alloys (SMAs). It was found that the shape memory effect of Cu-Zn-Al alloys was the best when the content of Gd was between 0.08wt.% and 0.12wt.%[8]. Luo Bing-hui found that 0.21wt.% Sc and 0.15wt.% Zr can improve the damping stability of Zn-22%Al[9]. Regarding magnesium and copper, the combined effects of two elements on zinc-aluminum alloys were often considered, Zn-Al-Cu-Mg alloys with medium Mg content (0.21wt.%) exhibited the highest yield strength at room and high temperatures, and ZnAl4Cu1-based alloys with less Mg showed excellent ductility at high temperatures[10]. However, there are few literatures on the effects of Mg/Cu alone, and their influence on the fracture mechanism of ZA27 alloy is not clear.

2. Experimental procedure

In this paper, the influence of Mg/Cu on the tensile properties of ZA27 alloy is explored by analyzing the microstructure of alloys with different compositions.

Zn-27Al-xMg and Zn-27Al-xCu alloys were prepared by the casting process. The copper and magnesium contents were designed to be 0.4wt.%, 0.8wt.%, 1.2wt.%, 1.6wt.%, and 2.0wt.%, respectively. Due to the poor performance of the as-cast alloy, the test as-cast alloy was annealed and solution treated. After consulting the literature[11～13], the specific heat treatment process was shown in Table 1.

Table 1. Heat treatment process of experimental sample

Number	ab	a1	a2	a3	a4	a5	b1	b2	b3	b4	b5
Alloy composition	Zn-27Al	Zn-27Al-xMg					Zn-27Al-xCu				
Heat treatment process	400 °C /h+125°C/24h	360 °C /h+120 °C /24h					460 °C /h+130 °C /24h				

ZEISS200MAT digital metallographic microscope, FEI QUANTA200F scanning electron microscope and X, Perth pro multi-functional X-ray diffractometer (XRD) were used to analyze the microstructure of materials. And Instron 4505 electronic universal testing machine was adopted to test tensile properties of materials

The material was cut into 10 mm×10 mm×5 mm with a metallographic sample cutter, and the samples were polished in the order of 150, 400, 800, 1200, 1500, and 2000 mesh sandpapers, then polished, and Cr_2O_3 was used as the polishing powder. The polished samples were etched with 4% nitric acid, and immediately rinsed with water after the appearance of grayish white corrosion producted on the surface , then wiped with alcohol and blew dry.

The tensile specimens are designed and processed according to the national standard GB/T6397-1986. The samples size processing standard was shown in Fig.1.

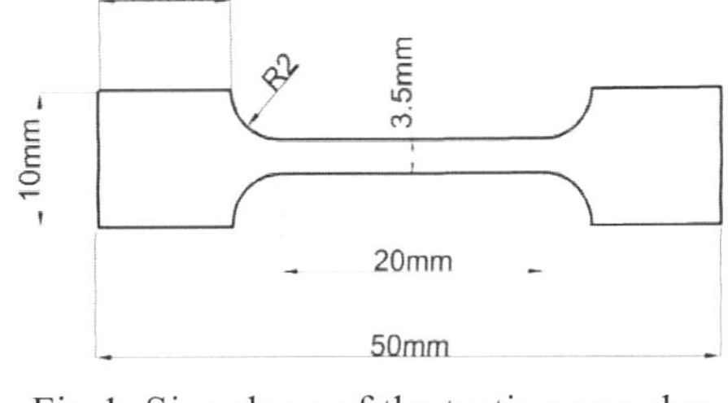

Fig.1. Size show of the testing samples

3 Microstructure analysis

3.1 Effect of Magnesium and Copper on Microstructure of ZA27 Alloy

According to the XRD test results of these alloys (Fig. 2), when magnesium was added to the ZA27 matrix alloy, Zn-27Al-xMg not only contained Al-rich phases η-Zn and eutectoid phases(α′+η), but also magnesium phases(Mg_xZn_y). However, copper and magnesium atomic radii were similar, indicating that copper remained in the matrix in the form of a replacement solid solution in Zn-27Al-xCu.

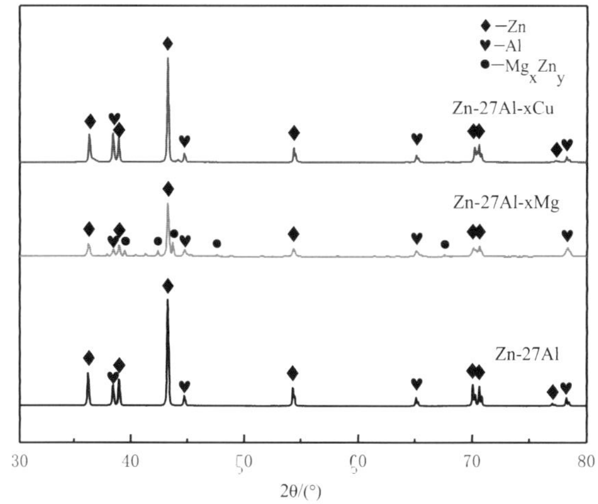

Fig.2. XRD test results of alloys

3.2 Effect of Magnesium Content on Tensile Properties of ZA27

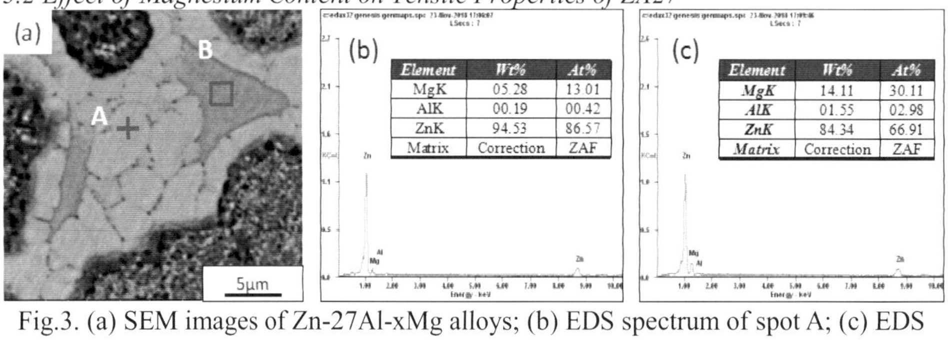

Fig.3. (a) SEM images of Zn-27Al-xMg alloys; (b) EDS spectrum of spot A; (c) EDS spectrum of spot B

According to the microstructure of Zn-27Al-xMg(x=0, 0.4, 0.8…2.0wt.%) after heat treatment, The microstructure was mainly composed of black η-Zn, gray black eutectoid phases(α′+η) and Mg phases(Mg_xZn_y). The bright white β (Mg_2Zn_3) phase was distributed in a network from Fig.3, and the gray β′ ($MgZn_2$) was continuously distributed in the gap of the β, and was coordinated with the β. With the increase of magnesium content, we can see that the amount of Mg phase was gradually increased and became more uniform from Fig.4. When the magnesium content reached 0.8wt.%, the Mg phases (β and β′) constituted a stable and uniform network structure. However, when the magnesium content continued to increase, the precipitation and growth of the Mg phases were inhibited, as η-Zn and (α′+η) continued to grow, they would be extruded into the surrounding Mg phases. And the network structure formed by the Mg phases became sparse.

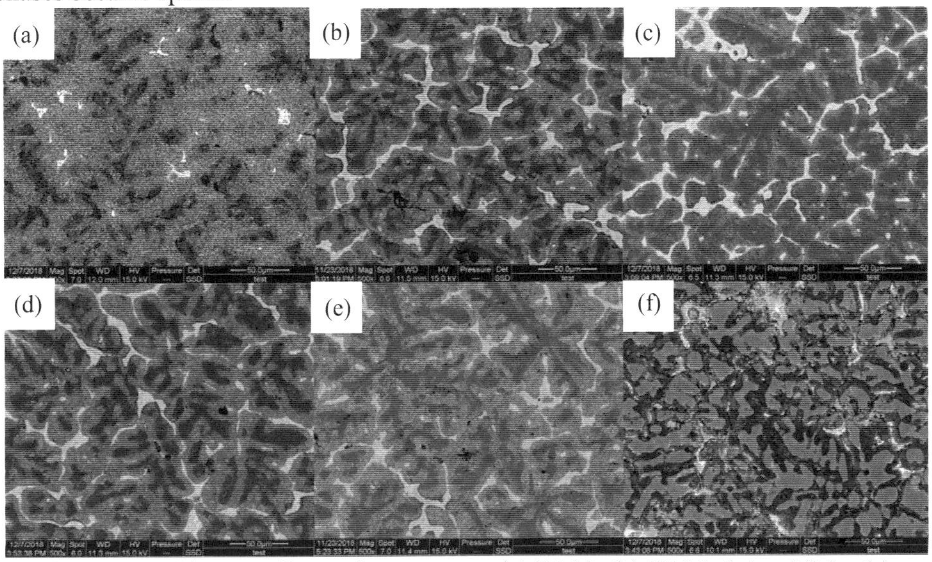

Fig.4. Zn-27Al-xMg alloys microstructure (a) ZA27; (b) ZA27−0.4wt.%Mg; (c) ZA27−0.8wt.%Mg; (d) ZA27−1.2wt.%Mg;(e) ZA27−1.6wt.%Mg; (f) ZA27−2.0wt.%Mg

Compared with the ZA27 binary alloy, the elongation of Zn-27Al-xMg(x=0.4, 0.8…2.0wt.%) was lower from the Table 2, and the general trend decreased with the increase of magnesium content, and the maximum was Zn-27Al-0.4Mg, which did not reach 15%. In the same series, the tensile strength of Zn-27Al-0.8Mg was the largest, which is 294.61MPa.

Table.2 Tensile properties of Zn-27Al-xMg(x=0, 0.4, 0.8…2.0wt.%)

Alloy composition	Percentage of magnesium in Zn-27Al-xMg(wt.%)					
	0	0.4	0.8	1.2	1.6	2
Tensile Strength(Mp)	175.96	188.22	294.61	240.97	190.59	180.96
Elongation(%)	14	11.6	10.55	9	5.6	6.4

As can be seen from Fig.5(a), there was no significant macroscopic plastic deformation near the crack, the surface had a relatively flush particle shape. Therefore, the fracture form of the Zn-27Al-xMg(x=0.4, 0.8…2.0wt.%) alloy was mainly brittle fracture.

Fig.5(b) showed the morphology of the Zn-27Al-xMg(x=0.4, 0.8…2.0wt.%) ductile fracture zone. Fig.4(c) showed the fracture surface morphology of the brittle dendrites in the final fracture zone. It can be seen from Fig.5(d) that the significant brittleness along the crystal plane fracture characteristics appeared at the interface between β and β′, indicating that the material did have intergranular brittleness. There was also a small amount of casting dimples between the magnesium phases, a fishbone pattern appeared from Fig.5(e), and the river pattern appeared on the fracture from Fig.5(f), which showed that this is a hybrid fracture mode dominated by brittle fracture.

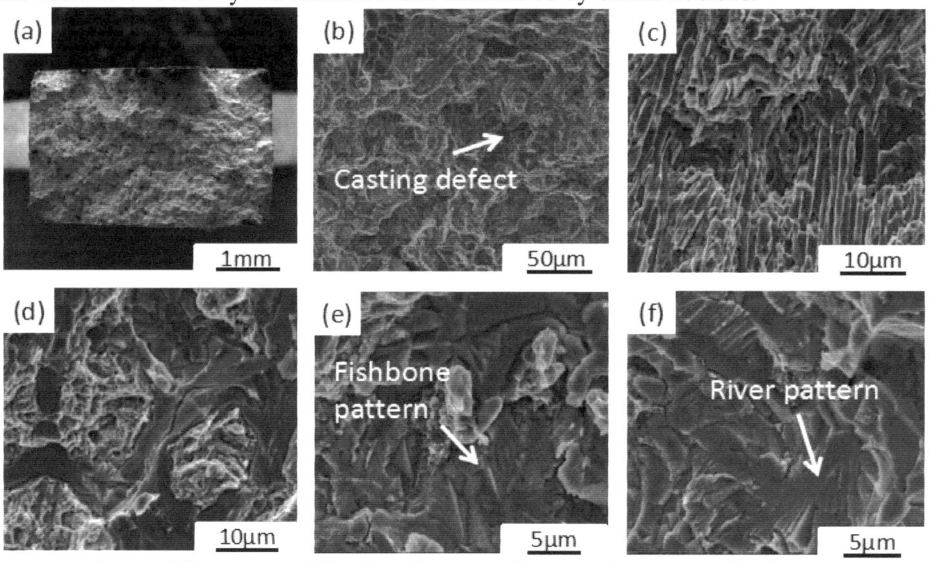

Fig.5. Different magnification diagram of Zn-27Al-xMg tensile fracture

According to the Cottrell dislocation reaction theory and the Smith theory, the fracture of the sample was related to the Mg phases(β and β′) precipitated in the eutectoid phases and the internal defects. On the one hand, since the Mg phases were hard phases and had a small angle sharp angle, It can be inferred from Fig.5 when the Mg content was less than 0.8wt.%, the Mg phases precipitate very little, which contained only dendrites, and were uniformly distributed at the grains boundary. When the Mg content was 0.8wt.%, the irregularly shaped Mg phases precipitated in a large amount, the size became larger, the brittleness increased, and the matrix was split. At the same time, the tensile deformation of the test alloy first occurred on the intergranular eutectoid phases or the crystal defects, resulting in local stress concentration at the tip of

IOP Conf. Series: Materials Science and Engineering **605** (2019) 012015 doi:10.1088/1757-899X/605/1/012015

the polygonal and irregularly shaped β′ during the stretching process, so that the β′ cracked or produced microcracks. However, since the Mg phases were arranged in a regular and stable network structure, the tensile strength of Zn-27Al-0.8Mg was large. On the other hand, the Mg phases were brittle phases, and as showed in Fig.5(d), there was a clear marked trace of brittle fracture on the Mg phases. Due to the presence of the gray Mg -rich phase β′ in the test alloy structure, lattice distortion easily occurred when an external force was applied, resulting in the material to be brittle fracture. Moreover, since the alloy was formed by casting, it was obvious that there were many crystal defects such as shrinkage holes in the fracture surface of Zn-27Al-xMg in Fig.5(b), which was also the main reason for the limited tensile strength of materials. The origins of the quasi-cleavage crack were the shrinkage cavity and other defects in the grain caused by casting and the crack source of the cleavage fracture is the grain boundary or the magnesium phase (β and β′) interface.

3.3 Effect of Copper Content Tensile Properties of ZA27 Alloy

It can be seen from Fig.6 that the microstructure of Zn-27Al-xCu is composed of black Zn-rich phase η-Zn and gray-black eutectoid phases (α′+η) from the SEM results. And black η distribution was relatively regular. When the copper content was 0.4wt.%, it was uniformly distributed in the form of needle or strip, and the eutectoid phases (α′+η) existed at the center of the η. As the copper content increased, black η phase became significantly smaller, gradually aggregated, and was distributed in the eutectoid structure in the form of dendrites or snowflakes (Fig.7). Even the number of grains had also increased and regional segregation occurred.

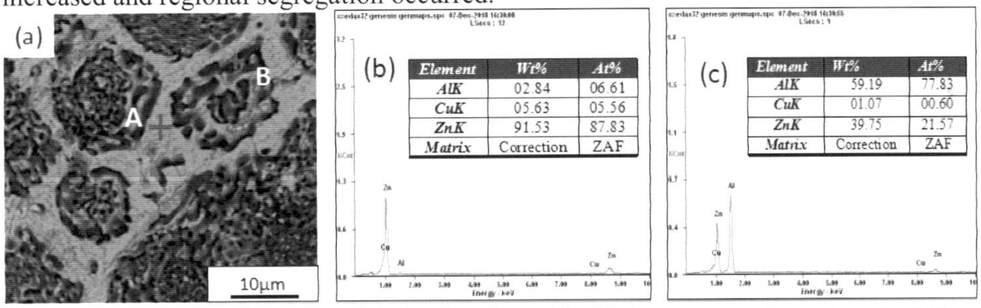

Fig.6. (a) SEM images of Zn-27Al-xCu alloy; (b) EDS spectrum of spot A; (c) EDS spectrum of spot B

It can be seen from Table 3 that the addition of copper can significantly improve the tensile properties and elongation of ZA27 alloy. The tensile strength of the alloy decreases and the elongation increases with the increase of copper content. In particular, the tensile strength of Zn-27Al-0.4Cu can reach 292.58MPa, and the elongation of Zn-27Al-2.0Cu reached 80%.

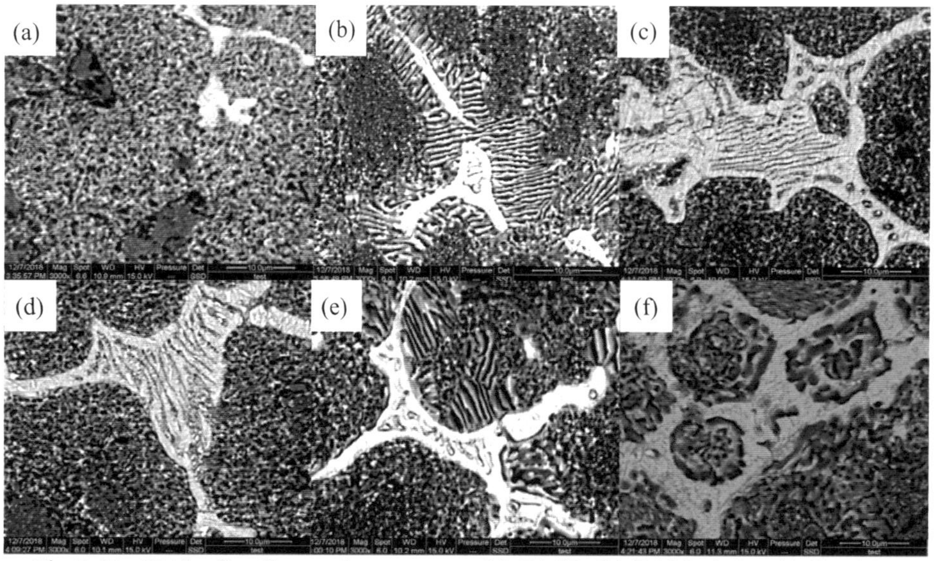

Fig.7. Zn-27Al-xCu alloys microstructure (a) ZA27; (b) ZA27−0.4wt.% Cu; (c) ZA27−0.8wt.% Cu; (d) ZA27−1.2wt.% Cu;(e) ZA27−1.6wt.% Cu; (f) ZA27−2.0wt.% Cu

Table.3 Tensile properties of Zn-27Al-xCu(x=0, 0.4, 0.8…2.0wt.%)

Alloy composition	Percentage of magnesium in Zn-27Al-xCu(wt.%)					
	0	0.4	0.8	1.2	1.6	2
Tensile Strength(Mpa)	175.96	292.58	267.79	264.08	264.55	242.8
Elongation(%)	14	20	40	40	65	80

As shown in Fig.8(a), the alloy had a large necking, and the appearance of the fracture was cup-shaped, and the bottom of the cup was perpendicular to the direction of the tensile stress. The surface of the fracture was goose-like, the color was gray, and it had obvious topographic characteristics of pure shear fracture. So the fracture form of the Zn-27Al-xCu alloy was ductile fracture.

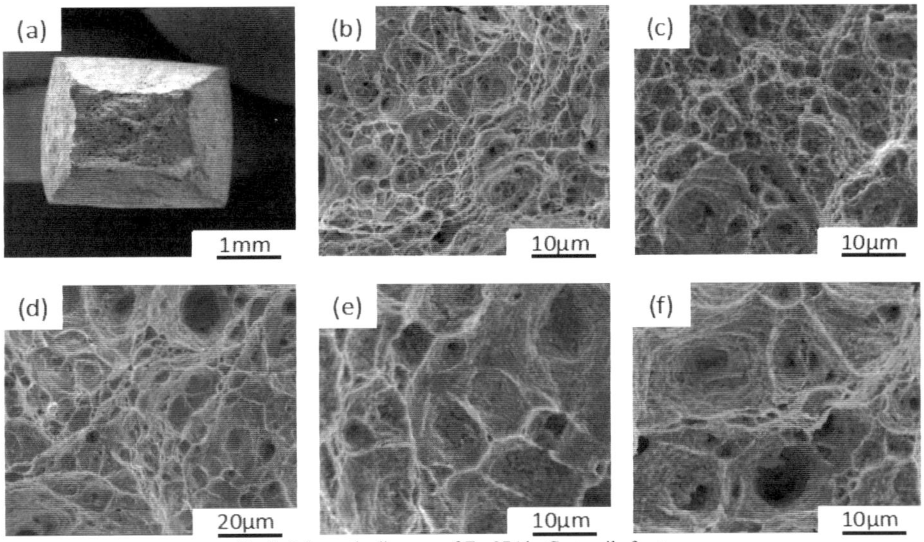

Fig.8. Schematic diagram of Zn-27Al-xCu tensile fracture

From Fig.8, the dimple morphology of the final fracture zone of the tensile fracture was observed, and second-phase particles appeared at the bottom of dimples. It can be seen from Fig.8 (a) that there were relatively porous holes in the material, and the holes were mainly divided into two parts. The first part is the casting defect and the second part is formed by the movement of the second phase particles (α'+η). The pores nucleated, grew up, and accumulated. Under the action of slip, the cavity gradually grew and was connected with other voids. The (α'+η) and the matrix η interface aggregated strength weaken. Therefore, with the increase of copper, more and more pores were formed by the sliding movement of the second phase particles, and the tensile strength of Zn-27Al-xCu is getting smaller and smaller.

4 Conclusions

In this paper, the effect of magnesium or copper and its content on the tensile properties of ZA27 alloy was studied. Through experiments and analysis, the following conclusions were obtained:

1. Magnesium was added to the ZA27 alloy, and magnesium phases (β and β') was formed; and the copper atom was present in the matrix as a replacement solid solution.
2. The tensile test results of Zn-27Al-xMg alloy showed that the fracture form was a mixed fracture mode dominated by brittle fracture. The magnesium phases (β and β') are the main reason for the brittle fracture of Zn-27Al-xMg alloy. With the increase of magnesium content, the tensile strength of Zn-27Al-xMg increased first and then decreased. The value of Zn-27Al-0.8Mg was the largest, which is 294.61MPa. However, due to the presence of brittle magnesium phase, the elongation of Zn-27Al-xMg was

smaller than the elongation of ZA27.

3. The fracture form of Zn-27Al-xCu alloy was ductile fracture, and the cause of the fracture was divided into two parts, one was related to the defects such as holes or inclusions generated during the casting process, and the other was related to the distribution, morphology and size of the eutectoid phase ($\alpha'+\eta$) . Therefore, with the increase of copper content, the tensile strength of Zn-27Al-xCu decreased. The tensile strength of Zn-27Al-0.4Cu was the largest, which was 292.58MPa.

Acknowledgements

This paper is funded by the International Exchange Program of Harbin Engineering University for Innovation-oriented Talents Cultivation.

References

[1] Liu C Y, Zhang X, Xia C Q, et al. 2017 *J. Mater. Sci. Eng. A.* 703:45-53.
[2] Rosenberger M R, Ares A E, Gatti I P, et al. 2010 *J. Wear.* 268(11):1533-1536.
[3] Lehuy H, G. L'espérance. 1991 *J. J. Mater. Sci.* 26(2):559-568.
[4] Zhu Y H, Torresvillasenor G, Pina C. 1994 *J. J. Mater. Sci.* 29(6):1549-1552.
[5] Zhu Y , To S , Liu X , et al. 2011 *J. Metall. Mater. Trans. A.* 42(7):1933-1940.
[6] Ren L B , Quan G F , Zhou M Y , et al. 2017 *J. Mater. Sci. Eng. A.* 690:195-207.
[7] Wan D, Wang J, Yang G. 2009 *J. Mater. Sci. Eng. A.* 517(1-2):114-117.
[8] Xu J W. 2008 *J. J. Alloys Compd.* 448(1-2):0-335.
[9] Wang J, Song P,F, Gao S, et al. 2011 *J. Mater. Sci. Eng. A.* 528(18):5914-5920.
[10] Wu Z, Sandl Bes S , Wu L , et al. 2016 *J. Mater. Sci. Eng. A.* 651:675-687.
[11] Lehuy H, G. L'espérance. 1991 *J. J. Mater. Sci. A.* 26(2):559-568.
[12] Lei H, Shuhong L, Yong D, et al. 2018 *J. Calphad,* 62:99-108.
[13] Kanayo A K, Shaibu U. 2018 *J. J. Mater. Sci. A.* S2468217918300340.

IOP Conf. Series: Materials Science and Engineering **605** (2019) 012016 doi:10.1088/1757-899X/605/1/012016

Study on wear behaviour of 4Cr4Mo2NiMnSiV die steel at high temperature

Renheng Han [a,b],Ning Li [a,b], Yao Huang [a,b], Zhiming Li [a,b], Ziming Bao [a,b], Chengzhi Zhao [a,b,*], and Hexin Zhang [a,b,*]

[a] College of Materials Science and Chemical Engineering, Harbin Engineering University, Harbin 150001, China

[b] Key Laboratory of Superlight Materials and Surface Technology, Ministry of Education

E-mail: zhaochengzhi@hrbeu.edu.cn, zhanghx@hrbeu.edu.cn

Abstract. In this paper, with reference to the composition and properties of widely used H13 hot working die steel, and the amount of alloying elements in the steel is reasonably improved, and a new steel grade of 4Cr4Mo2NiMnSiV die steel is designed. A new process was designed and the wear properties of 4Cr4Mo2NiMnSiV die steel were studied. SEM, EDS and XRD were utilized to observe the microstructure of the worn surface, and the wear products were analyzed to discuss the wear mechanism. Conclusion illustrates the coefficient of friction of cast steel 4Cr4Mo2NiMnSiV gradually increases with increasing temperature. Wear of the 4Cr4Mo2NiMnSiV die steel is mainly caused by spalling and high-temperature oxidation. The surface oxidation products of high temperature wear samples are Fe_2O_3 and Fe_3O_4, and spalling debris caused by fatigue cracks is blocky and a small amount of long flake oxide. The oxide film generated during the high temperature wear process effectively reduces the wear of the material. Through comparison experiments, it is found that the 4Cr4Mo2NiMnSiV die steel of martensite and bainite structure has the best wear resistance, followed by martensitic alloy steel, and the wear resistance of H13 steel which is involved in comparison is lower than that of alloy steel.

1. Introduction

Compared with the traditional hot forging mold, the casting mold can significantly improve the material utilization rate, shorten the manufacturing cycle, reduce the production cost, and facilitate the forming of complex molds. Moreover, cast hot work die steel has better red hardness, fatigue resistance and low notch sensitivity than forged steel. Therefore, casting forming hot work die has remarkable economy and advancement, and gradually becomes a new hot work die manufacturing technology [1-3]. However, hot die steel high content of alloying elements has a high content of alloying elements, and it tends to form a coarse structure and macrosegregation during solidification, resulting in early brittle fracture of the mold. Refining the grain by modification, reducing the segregation of alloying elements, and matching with reasonable heat treatment can significantly reduce the brittleness of the cast hot work die steel [4].

When the hot work die is working, the high temperature blank is deformed under the action of high load, and generates strong friction and wear with the surface of the cavity. High temperature wear is the main failure mode of hot work die and one of the key factors affecting the service life of hot work die. In recent years, material researchers have conducted extensive research on the high temperature wear behavior and wear mechanism of hot work molds. According to different physical processes during wear, Burwell divides wear into different types such as abrasive wear, adhesive wear, fatigue wear, erosion wear, corrosion wear and fretting wear, and thus established a foundation wear scientific research [5]. Quinn, as the proponent of the theory of spalling wear and oxidative wear, pointed out in his early research that the shedding of metal will accelerate the process of wear, and the shedding of the hot surface reaction layer will slow down the wear rate [6]. The reaction layer can be produced not only in the atmosphere, but also in the reaction with oil, additives and impurities to form a protective layer with protective properties [7]. Welsh pointed out in the study that the two most basic conditions for resisting severe wear are hardening and oxidation [8]. The study also shows that Fe_3O_4 protects steel better than Fe_2O_3 during wear, which results in low wear rate of steel [9]. The above research results show that, on the one hand, the microstructure and properties of the die steel affect its high temperature wear performance; on the other hand, the formation and shedding of the wear surface oxide film has an important influence on the high temperature friction and wear behavior of the hot work die steel. However, these studies are mainly for forging hot work die steel, and the high temperature wear of cast forming hot work die steel is less studied.

In this paper, 4Cr4Mo2NiMnSiV hot work die steel is used as the object to study the high temperature wear behavior of casting die steel at different temperatures, and the high temperature wear mechanism is discussed. The research results will be helpful for the optimization of casting hot work die steel and the die life prediction.

2. Experimental procedures

The experimental steel is obtained by high temperature melting of raw materials such as pig iron, electrolytic copper, low carbon ferrochrome, pure nickel, ferromolybdenum, vanadium iron ore, melting molten steel in GW medium frequency induction furnace. Table 1 shows the chemical composition of H13 steel and 4Cr4Mo2NiMnSiV steel. The heat treatment process of 4Cr4Mo2NiMnSiV steel is shown in Fig. 1. Among them, 4Cr4Mo2NiMnSiV steel, obtained by the process of (b), is the martensite structure, which is written as 4Cr4Mo2NiMnSiV (M), and the (c) is obtained by the martensite and bainite multiphase structure, which is written as 4Cr4Mo2NiMnSiV (M+B).

The abrasion test was carried out on a HT-1000 high temperature friction and wear tester. The test used a ball-and-disk type friction pair. The top of the grinding rod was a metal ball made of GCr15 bearing steel, and the grinding disc was a ϕ30 mm×10 mm disc made of the tested material. The experimental conditions are dry sliding friction and wear conditions. The experimental load was 0.5 Kg. The grinding wheel has a rotation frequency of 10 Hz. The friction time is 30 min. The total taxi distance d is approximately 3.4×102 m. The experimental temperature was set to room temperature, 200°C, 400°C, and 600°C. During the experiment, the electric resistance furnace is used for automatic heating, and the temperature is controlled by thermocouple and temperature control instrument. The friction coefficient, wear amount and wear rate are directly measured by the computer. After the abrasion test, the phase composition of the material was analyzed by X Perth pro type multifunctional X-ray diffractometer (XRD) (Cu target, scanning range 2θ was set to 10-110°, scanning speed was 3°/min). The FEI QUANTA 200F scanning electron microscope was used to observe the surface morphology after wear, and the fracture mechanism, oxidation and wear characteristics of the material were analyzed and judged. The energy dispersive X-ray spectrometer (EDAX) attached to the scanning electron microscope was used to analyze the energy spectrum of the material.

Table 1 Chemical compositions of H13 and 4Cr4Mo2NiMnSiV steel (mass fraction/%).

	C	Cr	Mo	V	Si	Mn	Ni	P, S
H13	0.38	5.15	1.45	0.90	1.04	0.43	/	<0.03
4Cr4Mo2NiMnSiV	0.40	4.00	2.30	0.60	0.40	0.60	1.00	<0.03

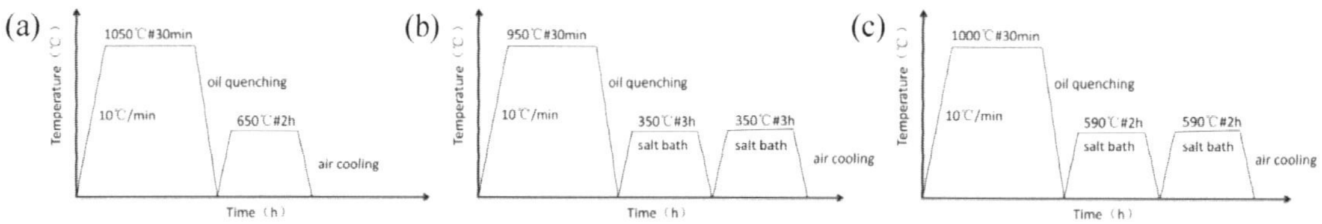

Fig.1. Heat treatment process flow chart: (a) H13 steel; (b) 4Cr4Mo2NiMnSiV (M) steel; (c) 4Cr4Mo2NiMnSiV (M+B) steel

3. Results and Discussions

3.1. *Coefficient of friction*

Fig. 2 shows the coefficient of friction of 4Cr4Mo2NiMnSiV (M) steel with time at different temperatures. The results show that the friction of each sample has no obvious running-off period at different temperatures, and the friction coefficient reaches a stable value in a short time. At room temperature, the coefficient of friction is the lowest, but the coefficient of friction at room temperature fluctuates relatively more time. Studies [7] have shown that when the wear is severely worn, the contact between the two surfaces may cause plastic transfer from one surface to the other, resulting in wear debris. Plastic flow in the transfer phase is an important factor in the large fluctuation of the friction coefficient. As the temperature increases, the coefficient of friction also increase due to the coefficient of friction of the material is related to the state of the contact surface. When the friction surface of contact has a large rigidity, and the lower roughness has no direct interaction with the grinding metal, the friction coefficient of the material is small. On the contrary, the friction coefficient of the material is relatively high, and the temperature has a great influence on these factors. However, as the temperature increases, the fluctuation of the friction coefficient decreases. This is because in the process of high temperature wear, as the ambient temperature increases, the material will temper soften, resulting in an increase in the contact area of the friction surface. As the temperature increases, the thickness of the surface oxide film gradually increases. The effect of the oxide film on the friction is reflected in the following two aspects. On the one hand, the oxide film can reduce the frictional force when the surface microprotrusions are in contact. On the other hand, the oxide is low in strength and easily deformed to reduce friction [10]. And the temperature rise increases the degree of oxidation of the surface of the 4Cr4Mo2NiMnSiV (M) steel, and the appearance of the oxide makes the friction surface rough and increases the friction coefficient.

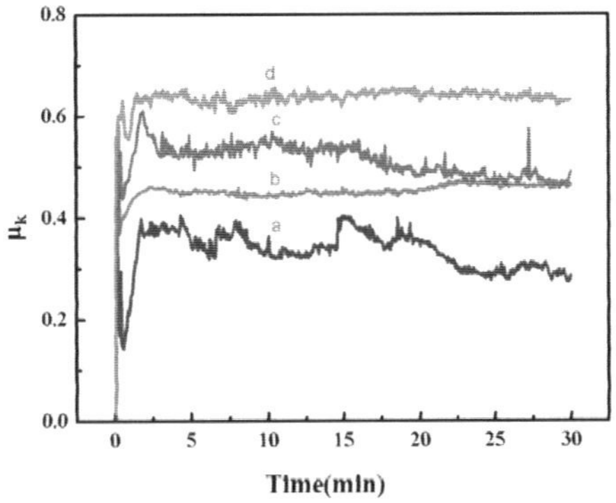

Fig. 2. 4Cr4Mo2NiMnSiV (M) friction coefficient at different ambient temperatures (μ_k: friction coefficient): (a) Room temperature; (b) 200℃; (c) 400℃, (d) 600℃

Fig. 3 shows the relationship between the average friction coefficient and the wear rate of 4Cr4Mo2NiMnSiV (M) steel. The wear rate of 4Cr4Mo2NiMnSiV (M) steel increased when the temperature was raised to 200 °C, and the wear resistance of the material decreased. When the temperature rises above 400 °C, the wear rate of 4Cr4Mo2NiMnSiV (M) steel decreases again. In the process of high temperature wear, due to the change of ambient temperature and the generation of heat in the friction engineering, an oxide layer appears on the worn surface, and the appearance of the oxide layer and covering the surface of the substrate will protect the material, which can make the wear rate reduce. However, most of the oxides are brittle, and peeling occurs under the action of friction, and the wear rate is thus increased. The rate of exfoliation of the oxide generally depends on the amount of adhesion between the oxide and the substrate and the support of the substrate to the oxide. The wear rate of 4Cr4Mo2NiMnSiV (M) steel increases from room temperature to 200 °C because the generated brittle oxide is peeled off under the action of friction. When the temperature is raised to 200 °C, the oxide formed has a strong adhesion to the substrate, which can effectively protect the substrate, and thus the wear rate is lowered.

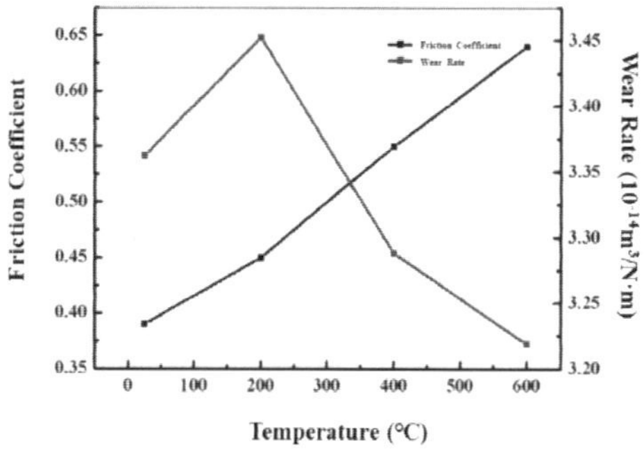

Fig. 3. Wear rate and average friction coefficient of 4Cr4Mo2NiMnSiV (M) steel at different temperatures

3.2. Wear surface

Fig. 4 shows the wear surface morphology of H13 steel, 4Cr4Mo2NiMnSiV (M) steel and 4Cr4Mo2NiMnSiV (M+B) steel at 600 °C. The wear rate measured by the experiment are 4.012, 3.589 and 3.216 (10^{-14} m^3/N·m). It can be seen from the SEM pictures in Fig. 4(a) and (d) that there are deep pits in the peeling zone of H13 steel, and the surface wear and peeling of H13 steel under the same experimental conditions is the most serious. The wear surface is oxidized to a deeper extent, with blocky and granular wear debris, fine scratches locally, less oxide around the scratches, and obvious cracks at the edges, which are characterized by adhesive wear and oxidative wear. It can also be found in Fig.4 (b) and (e) that although the surface of 4Cr4Mo2NiMnSiV (M) steel is not severely peeled off, there are obvious furrows and granular wear debris along the rubbing direction, wear depth and peeling pit. The area is large, showing plow wear and abrasive wear. Figure 4(e) also shows that the depth of the spalling zone is also deeper than that of 4Cr4Mo2NiMnSiV (M+B) steel, which is also reflected in the difference in wear rate of the three samples. From Fig. 4(c), 4Cr4Mo2NiMnSiV (M+B) steel is distributed with intermittent strip-shaped wear scars, and block-shaped abrasive chips are distributed on both sides, and a large number of cracks from the outside to the inside are generated on the surface of the wear scar. As the cracks expand, they form new wear debris. And there are a large number of fine wear oxide particles in the peeling zone of the three samples.

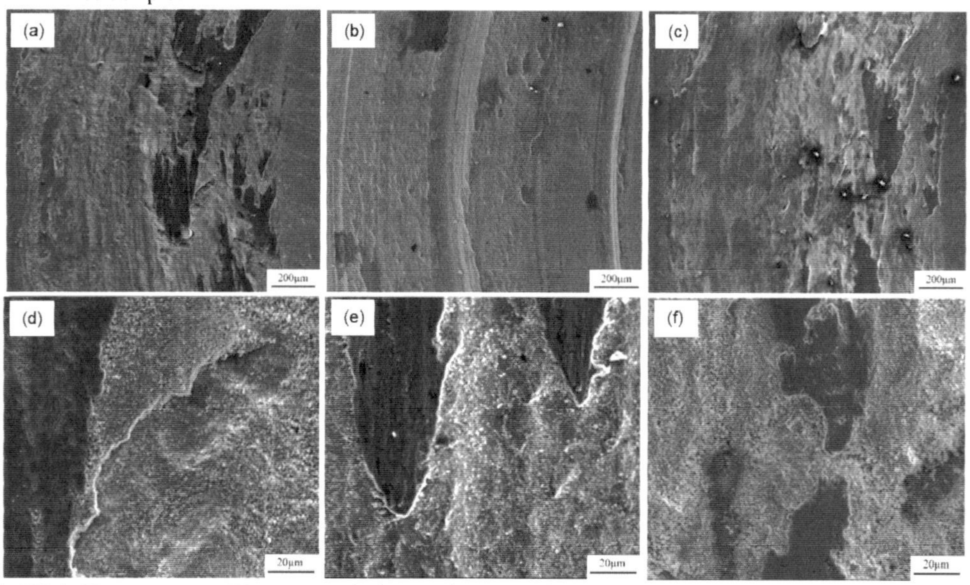

Fig. 4. High Temperature Wear Morphology of H13 and 4Cr4Mo2NiMnSiV steel at 600 °C (SEM): (a) (b) H13 steel; (c) (d) 4Cr4Mo2NiMnSiV (M) steel; (e) (f) 4Cr4Mo2NiMnSiV (M+B) steel

The morphology and composition of the wear debris of H13 steel and cast 4Cr4Mo2NiMnSiV specimens at 600°C were observed by SEM-EDS (Fig.5). The results showed that the shape of the abrasive chips was blocky (containing a small amount of long flakes) and granular, and the average oxygen content in the wear debris was about 34.74%. Combining the analysis of the wear surface (Fig.4) and the cross section (Fig.7), the high temperature wear process of the cast 4Cr4Mo2NiMnSiV steel is known. As the friction progresses, the surface of the sample oxidizes to form an oxide layer. Due to frictional heat generation, the wear scar position is oxidized faster than both sides. Under the action of fatigue stress and internal stress of the oxide, cracks are generated on both sides of the wear scar and spread to the periphery. When the crack propagates to the surface of the substrate, the position of the contact surface between the oxide layer and the substrate gradually oxidizes, which makes the combination of the two unstable. When cracks on both sides penetrate, large-sized block-shaped abrasive chips are formed, and some of the block-shaped abrasive chips are crushed and become fine particles of abrasive debris. It is concluded that the main mechanism of high temperature wear of cast 4Cr4Mo2NiMnSiV sample steel is oxidative spalling wear and fatigue wear. Due to the lowest content of Fe in H13 steel, the loss of Fe in the matrix during wear is the most. Therefore, the 4Cr4Mo2NiMnSiV sample steel has better resistance to wear than H13 steel, which also has a good correspondence with the wear rate of the material.

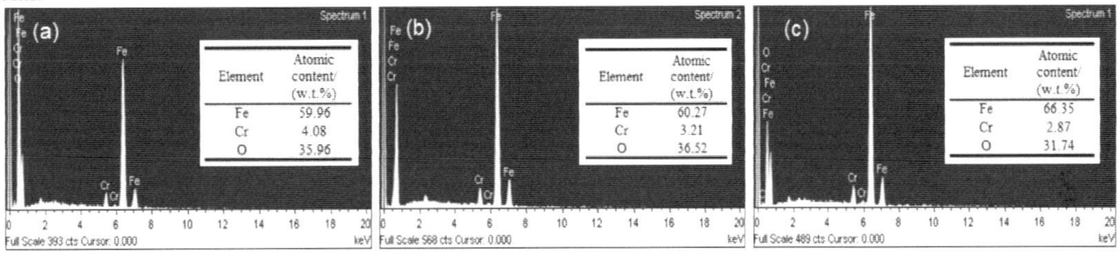

Fig. 5. EDS Analysis of Wear Surface of H13 and 4Cr4Mo2NiMnSiV steel at 600 °C: (a) H13 steel; (b) 4Cr4Mo2NiMnSiV (M) steel; (c) 4Cr4Mo2NiMnSiV (M+B) steel

3.3. Analysis of the wear surface phase

Fig. 6 shows the XRD phase analysis of the worn surface of H13 steel and cast 4Cr4Mo2NiMnSiV specimen at 600 °C. From the figure, the wear surface of the three samples after the high temperature wear at 600 °C showed the oxide of Fe, and the surface also had the presence of matrix iron, indicating that the wear surface did not completely oxidize and peel off during the wear process. The oxide of the cast 4Cr4Mo2NiMnSiV sample is mainly composed of Fe_2O_3 and Fe_3O_4, and its peak intensity is obviously higher than that of α-Fe, which indirectly indicates that a uniform oxide layer is formed on the worn surface and is consistent with the surface morphology analysis results.

Generally, Fe oxides include FeO, Fe_3O_4, and Fe_2O_3. FeO generates temperatures above 600 °C and Fe_2O_3 forms temperatures from 200 °C to 600 °C [10-11]. The flash point calculation formula based on Arachard [12]:

$$(\theta_x)_{max} = 1.64\ \theta_m \tag{1}$$

Where $(\theta_x)_{max}$ is the maximum temperature of the friction surface, and θ_m is the ambient temperature. When the temperature is 600 °C, the flash point temperature of the worn surface is much higher than 600 °C, but the FeO phase is not detected by XRD. Due to the high temperature friction, the surface defects of the material increase, which is favorable for the diffusion of Fe atoms and O atoms and forms high oxygen content Fe_3O_4 and Fe_2O_3. Comparing the XRD results of different samples, it can be found that the oxide film after wear at 600 °C is mainly Fe_2O_3. For the 4Cr4Mo2NiMnSiV steel, Fe_3O_4 is also present in the oxide film. Oxidation spalling is the primary cause of material wear during high temperature friction, and a stable oxide layer provides effective protection for the substrate. The existing research results show that both Fe_2O_3 and Fe_3O_4 oxides have antifriction effect, but Fe_3O_4 antifriction effect is better than Fe_2O_3 [13]. Thereby reducing the wear rate of the material, which is also an important factor in the wear resistance of alloy steel under the same conditions is higher than that of H13 steel.

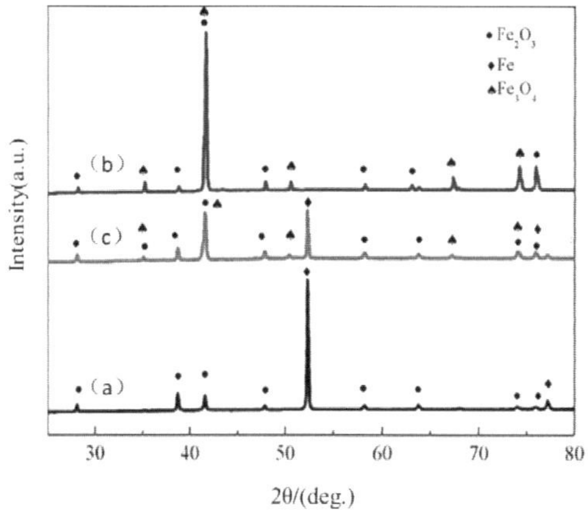

Fig. 6. XRD pattern of wear surface of H13 and 4Cr4Mo2NiMnSiV steel: (a) H13 steel; (b) 4Cr4Mo2NiMnSiV (M) steel; (c) 4Cr4Mo2NiMnSiV (M+B) steel

3.4 Wear section

Fig.7 shows the wear profile and EDS analysis of H13 steel and cast 4Cr4Mo2NiMnSiV steel after oxidation at 600 °C for 200 h. The three samples have a thick oxide layer on the surface layer, and a layered inner oxide layer and an unoxidized layer exist in the lower portion of the oxide layer. The inner layer is oxidized from inclusion and microspores. During the friction process, oxygen atoms easily enter the matrix through the micropores or microcrack defects of the material matrix, causing oxidation inside the matrix. When the surface oxide layer is detached, the inner flaky oxidized region forms an oxide layer, and a region having a low oxygen content is formed at the center of the high oxygen content oxide layer. This area has weak bonding ability and is easy to crack and peel off. Thus reducing the wear resistance of the material. After calculation, the thickness of the oxide layer of H13 steel, 4Cr4Mo2NiMnSiV (M) steel and 4Cr4Mo2NiMnSiV (M+B) steel is 58.27μm, 89.23μm and 97.62μm. H13 steel oxide layer is the thinnest, divided into two layers inside and outside, the outer layer is loose, small holes appear in some positions, the inner layer is dense, and the matrix is firmly bonded, and there is no crack in the scanned photograph. The results of EDS analysis showed that the outer oxide layer O aggregated. The Fe content was slightly lower than that of the matrix, and the internal oxide layer had more Cr elements. 4Cr4Mo2NiMnSiV (M) steel oxide layer thickness is large, also divided into two layers inside and outside. The outer layer is loose and porous, and the void is larger and larger than H13 steel. After EDS analysis, the content of Fe and O is higher than that of the matrix, and the main component is presumed to be the oxidation of Fe. The content of the inner Cr and Ni elements is enriched with respect to the matrix, so that a relatively dense Cr-Ni oxide layer is formed, which can effectively prevent the oxidation reaction from continuing. The 4Cr4Mo2NiMnSiV (M+B) steel oxide layer is also clearly divided into a tight inner oxide layer and a loose outer oxide layer, but it can be seen that the oxide layer thickness is the largest among the three samples. The results of EDS show that the content of Fe and O in the outer layer is higher than that in the matrix, and the inner layer Cr and O are segregated. It is speculated that the outer layer is the oxide of Fe and the inner layer is the oxide of Cr. During repeated enthalpy of external force on the surface of the sample, the source of the crack is reduced and the tendency of oxidation inside the substrate is suppressed. The thick oxide layer produced by the 4Cr4Mo2NiMnSiV die steel during the wear process provides stable support and resistance to plastic deformation and thermal softening. Therefore, the 4Cr4Mo2NiMnSiV die steel has better wear resistance at high temperature.

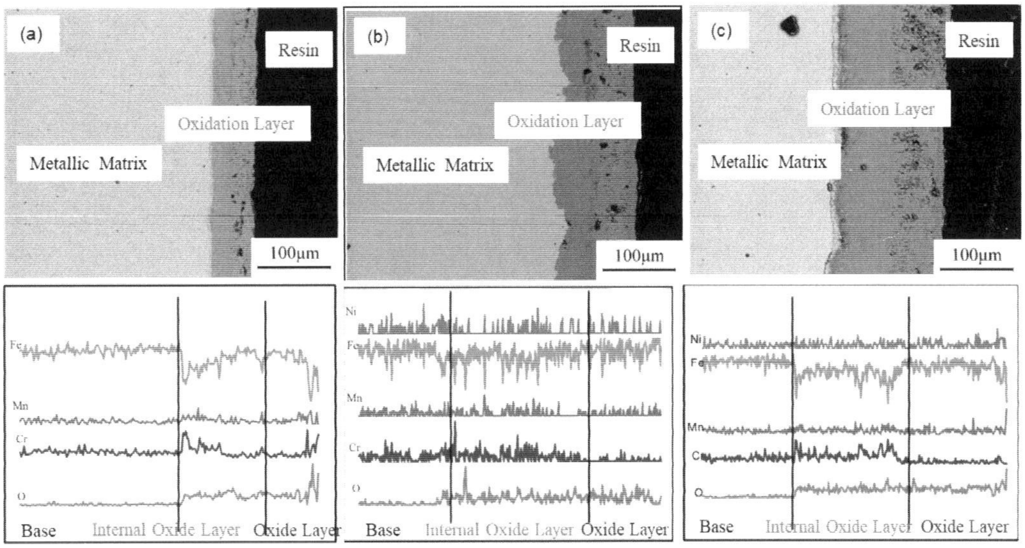

Fig. 7. Comparison between the simulation and experimental true stress-strain curves at different temperatures and strain rates: (a) 0.01 s^{-1}, (b) 0.1 s^{-1}, (c) 1 s^{-1}.

4. Conclusions

In this paper, the properties of the high temperature wear mechanism of 4Cr4Mo2NiMnSiV hot work die steel were studied. Through experiments and analysis, the following conclusions were obtained:
1. The friction coefficient of cast 4Cr4Mo2NiMnSiV steel increases with the increase of temperature, and the main mechanism of high temperature wear is oxidative spalling wear and fatigue wear.
2. Surface oxidations of cast 4Cr4Mo2NiMnSiV steel are Fe_2O_3 and Fe_3O_4. The oxide film generated during the high temperature wear process effectively reduces the wear of the material.
3. The 4Cr4Mo2NiMnSiV (M+B) steel has the best wear resistance, followed by 4Cr4Mo2NiMnSiV (M) steel, and the wear resistance of H13 steel is lower than that of 4Cr4Mo2NiMnSiV steel.

Acknowledgements

Project is supported by the Fundamental Research Funds for the Central Universities of Ministry of Education of China (Grant No. GK2100260214) (Grant No. GK2030260169), Equipment Pre-research National Defense Science and Technology Key Laboratory Fund (Grant No. 61429030101051704) and Department of Physics and Materials Sciences, Chiang Mai University. This paper is funded by the International Exchange Program of Harbin Engineering University for Innovation-oriented Talents Cultivation.

References

[1] Wang Shuqi, Wang Feng, Chen KM, et al. 2007*J. ISIJ Int.* 47(9): 1335-1340.
[2] Liu Jinhai, Li Guolu, Hao Xiaoyan, et al. 2005*J. China Foundry.* 2(4): 260-263.
[3] Drs. David Schwam, John F. Wallace. 2015*J. Die. Cast. Eng.* 26 (2): 12-16.
[4] Liu Hanghang, Fu Paixian, Liu Hongwei, et al. 2017*J. Metals.* 7:435-436.
[5] Bajguirani , H R H. 2016*J. Mater. Sci. Eng. A.* 338: 142-159.
[6] J K Lancaster. 2012*J. Proc. Roy. Soc.* A273: 466-483.
[7] M. Kerridge, K. Lancaster. 2014*J. Proc. Roy. Soc.* A236: 250-264.
[8] Quinn. 2008*J. Wear.* 216: 262-275.
[9] E H Lloeser, D Tabor, B B Twiss. 2012*J. ASLE Trans.* (13): 199-207.
[10] Molgaard J, Srivastava V K. 1977*J. Wear.* 41(2): 263-270.

IOP Conf. Series: Materials Science and Engineering **605** (2019) 012016 doi:10.1088/1757-899X/605/1/012016

[11] Beeley P R, Baker A J, Blackmore P A, et al. 2013*J. Met. Sci. J.* 8(1): 268-273.
[12] Archard J F, Rowntre R A. 1988*J. Wear*. 128: 1-17.
[13] Sullivan J L, Athwal S S. 1983*J. Tribol. Int.* 16(3): 123-131.

Effects of heat treatment on microstrucutures of a Re-containing Ni-based single crystal superalloy

J Zhang[1], L Q Gao[1,2,*], Y F Liu[3], J X Zhao[1] and X T Zhang[1]

1. Science and Technology on Advanced High Temperature Structural Materials Laboratory, Beijing Institute of Aeronautical Materials, Beijing 100095, P. R. China
2. School of Mechanics, Civil Engineering and Architecture, Northwestern Polytechnical University, Xi'an, 710072, P. R. China
3. AECC Sichuan Gas Turbine Establishment, Chengdu 610500, P. R. China

E-mail: zhangjian620@foxmail.com; lqgao0214@163.com; 824300948@qq.com; 455531944@qq.com; www.529881022@qq.com.

* Corresponding Author: lqgao0214@163.com

Abstract. Effects of heat treatment on microstrucutures of a Re-containing single crystal superalloy have been investigated. The results showed that (γ+γ') eutectic in as-cast alloy is practically dissolved with solution heat treatment, and the degree of segregation is distinctly decreased. The size, morphology and distribution of γ' phase are mightily affected by the first-stage ageing treatment temperature. When the first-stage ageing treatment temperature lower than 1160℃, the shape of the γ' phase is cubic. But the γ' phase shape is transformed to sphericity when the temperature higher than 1180℃. The γ phase channel width is significantly increased, and the fine γ' phase precipitated in the γ phase matrix channel.
Key words. Superalloy; Heat treatment; Microstrucutures; γ phase matrix channel

1. Introduction

Nickel-based single crystal superalloy is the key material for advanced aeroengine turbine blades and ground gas turbine blades. Monocrystalline alloys are prepared by selective crystallization or seed crystallization growing in dendrite orientation. Due to solute redistribution during the solidification process, a large amount of Al, Ta and other elements forming the γ' phase will be enriched in the dendrite cores of as-cast alloys, while Re, W, Mo and other elements forming the γ phase will be enriched in the dendrite cores [1,2]. The solidification segregation of cast superalloy results in the low degree of alloying in some parts of the alloys, which decreases the properties of the alloys; while the high degree of alloying in other parts promotes the precipitation of TCP phase and reduces the properties of the alloys [3]. Meanwhile, during the solidification and cooling process of single crystal superalloy, coarser and inhomogeneous γ' phases are precipitated, and there are (γ+γ') eutectics in the interdendritic region, which can also significantly reduce the mechanical properties of the alloys [4].

The purpose of heat treatment of nickel-based single crystal superalloy is to reduce dendritic segregation during the directional solidification process, to adjust the morphology and size of the γ' phase in single crystal superalloy, and to improve the mechanical properties of the alloys. Heat treatment process usually includes solution treatment and aging treatment. The main function of solution treatment is to solute all or most of the coarse γ' phase and γ/γ' eutectic in as-cast alloys, thereby improving the homogeneity of the structure and reducing the solidification segregation [5]. In order to further develop the performance potential of single crystal alloys, two-stage aging heat treatment is usually adopted after solution heat treatment to improve the properties of the alloys. It is generally believed that the first ageing treatment is a process of growth of γ' phase and the second ageing treatment is a process of increasing the square degree of γ' phase [6].

Persistent creep rupture is one of the most important mechanical properties of nickel-based single crystal alloys, and has been extensively investigated. The γ' phase is the most important strengthening phase, and its shape, size, volume fraction and distribution determine the creep rupture properties of the alloy [7]. At present, there are still many disputes about the type of organization is beneficial to creep properties and how the organization affects creep properties. And P. Caron and T. Khan [8] studied CMSX-2 and found that the best creep properties of the alloy can be obtained by choosing appropriate ageing treatment process, so that the γ' phase in the alloy eventually becomes a regular square with a size of about 0.45μm. However, Mackay and Ebert [9] observed different situations in Alloy 143, and the creep properties of the alloy are improved when the size is refined from 0.44μm to 0.15μm. At the same time, Murakumo et al. [10] studied that as the main strengthening phase, the optimum morphology of γ' phase in unequal alloys is different, so it is necessary to study each alloy in detail.

By adding Re and other refractory elements in the second and third generation single crystal alloys, the creep rupture properties of the alloys are significantly improved [2]. However, Re is a strong negative segregation element and promotes the segregation of other elements, resulting in the decrease of the uniformity of the alloy structure. Therefore, the research on heat treatment process of single crystal alloys containing Re becomes more and more vital. The heat treatment of single crystal superalloy has been deeply studied in the United States and Russia, and the precipitation dissolution and the number and distribution of strengthening phases in the alloys have been systematically researched, and different heat treatment systems are recommended for the same alloy to adapt the requirements of different working conditions [11,12]. At present, the research on the heat treatment process of single crystal superalloy is not systematic in China, especially the research of the content, size, morphology and homogeneity of the strengthened phase on the rupture properties of the alloys containing Re is not much.

DD11 alloy is the second-generation nickel-based single crystal superalloy containing Re developed independently by Beijing Institute of Aeronautical Materials in recent years, and its content of Re is 3 wt%, and it is mainly used to manufacture advanced aeroengine turbine blades [1]. In this research, the content, size and morphology of the strengthened phase of the alloys under distinct ageing heat treatment conditions are investigated, and the relationship between microstrucutures and temperatures is discussed.

2. Experimental materials and methods

In order to study the effect of heat treatment on the structure and rupture properties of nickel-based single crystal superalloy, the DD11 alloy developed by Beijing Institute of Aeronautical Materials is chosen as the base alloy, and its chemical composition (wt,%) is presented in Table 1. In HRS directional solidification furnace, single crystal rods with [001] orientation of 15 mm in diameter and 150 mm in length are prepared by Spiral crystal selection, and the pulling speed is 3 mm/min during directional solidification. The orientation of single crystal is determined by backscattering Laue method, and the rod whose growth direction deviated from [001] direction is less than 10 degrees is selected as the experimental material. On the NETZSCH STA449C differential thermal analyzer (DSC), the phase transformation temperature of the alloy is measured at a rate of 10℃/min from 1000℃ to 1430℃. According to the principle of eliminating eutectic, reducing dendrite segregation and avoiding primary melting, the solid solution heat treatment system is determined. Under this system, six different temperatures are selected for the first-stage ageing treatment. Then the same second-stage aging treatment is carried out for the test sample. The first-stage aging treatment temperature is 1100℃,1120℃,1140℃,1160℃,1180℃ and 1200℃, and after holding for 4 hours, air-cooling is carried out. The second-stage aging treatment is air cooling after holding for 32 hours at 870℃ (Table 2). After different heat treatments, the alloy is processed into standard specimens with gauge diameter of 5 mm, gauge length of 25 mm and total length of 66 mm. The high temperature endurance test is carried out at 1100℃/140 MPa and 980℃/250 MPa, and the endurance life is determined by the average values of three specimens. The corrosive solution of metallographic specimens is 1% HF + 33% HNO_3 + 33% CH_3COOH + 33% H_2O (vol.%). Microstructure morphologies at low and high power are observed by using optical microscopy (OM) and secondary electron mode of SUPRA 55 field emission scanning electron microscopy (FE-SEM). In this work, the volume fraction of γ' phase is measured by point analysis method, and the size of γ' phase is measured by Image-Pro software. The same specimens are counted by 6 to 8 photos.

Table 1. Chemical composition of the DD11 alloy (Wt. %).

Element	Al	Cr	Ta	W	Co	Mo	Re	C	Hf	Ni
Mass fraction/%	6.0	4.0	7.0	7.0	8.0	2.0	3.0	0.01	0.15	surplus

The elemental solidification segregation coefficient S_i and the solidification segregation coefficient S_i of as-cast alloy after solution heat treatment are measured by JEOL JXA-8800R electron probe (EPMA).

Elemental solidification segregation coefficient S_i can be expressed as:

$$S_i = C^i_{dendrite} / C^i_{interdendrite} \qquad (1)$$

In the formula, the sum is the concentration of element i in the dendrite cores and between the interdendrites. $S_i > 1$ indicates that the alloy element i segregates to the dendrite cores, while $S_i < 1$ indicates that the alloy element i segregates to the interdendrites.

Table 2. Solid Solution and Aging Heat Treatment Systems of Experimental Alloys.

Specimens	Solution treatment	First-stage aging treatment	Second-stage aging treatment
A		1100 ℃/4h, AC	
B	(1290/2h+	1120 ℃/4h, AC	
C	1300℃/2h	1140 ℃/4h, AC	870 ℃/32 h, AC
D	+1310℃/2h)	1160 ℃/4h, AC	
E	+1320℃/6h	1180 ℃/4h, AC	
F		1200 ℃/4h, AC	

3. Experimental results

3.1. Microstructure and transformation temperature of as-cast alloy

Fig.1 (a) is Optical microscopy image of as-cast alloy. It can be seen from the graph that the structure of the alloy presents typical dendrite morphology The primary dendrite spacing is 340μm. The white brightness contrast between interdendrites of the alloy is γ/γ' eutectic structure, and the volume fraction of the eutectic is about 6%. Fig.1 (b) and Fig.1 (c) are morphologies of dendrite cores and interdendritic morphologies of as-cast alloys respectively. It can be seen that the dendrite cores γ' phase are all butterfly-shaped with uneven size and great difference. In addition, the γ' phase is relatively small and the average size of the γ' phase is about 400nm. The interdendritic γ' phase size is larger, the cubic degree is better than the dendrite cores and the average size of the γ' phase is about 720nm.

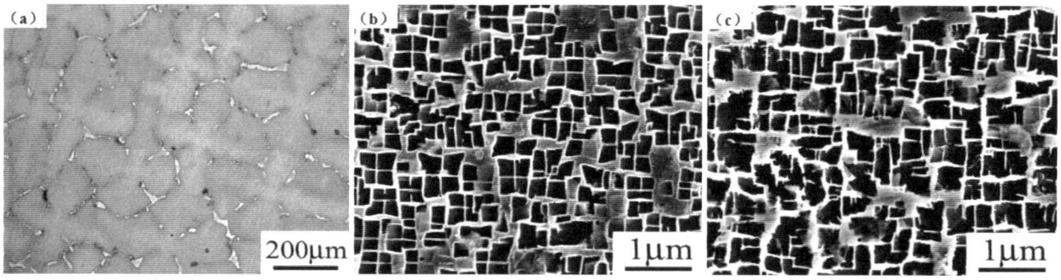

Fig. 1. Microstructures of as-cast alloys: (a) Optical microscopy image; (b) SEM image of dendrite cores; (c) SEM image of interdendrites.

Fig. 2. shows the DSC heating curve of as-cast alloys. There are three endothermic reactions on DSC heating curve. The first endothermic peak is due to the solid solution of the γ' phase between dendrite cores and interdendrites, and the dissolution temperature of the twisted phase of the alloy is about 1250℃. The second endothermic peak is the melting temperature of γ/γ' eutectic. The eutectic dissolution temperature measured by DSC is about 1351℃. The third endothermic peak is the liquidus of the alloy, which is about 1404℃. At the same time, the solid phase line is the intersection temperature of the endothermic peak tangent and the baseline, and the solid phase line of the alloy is about 1368℃.

In DSC measurements, a small amount of liquids did not fully solidify in a wide temperature range during the late solidification stage, and the thermal effect of these phase transitions is very weak. Therefore, there is a certain deviation between the phase transformation temperature measured by DSC curve and the actual phase transformation temperature of the alloy. In order to determine the actual "heat treatment window", under the guidance of the results of differential thermal analysis, the rapid water-cooled structure of the alloy sample after holding at different temperatures (1300℃~1350℃) is observed, that is, the eutectic dissolution temperature and initial melting temperature of the alloy are determined by metallographic method, and a reasonable solution treatment system is finally worked out. Fig. 3. shows the variation of volume fraction of eutectic dissolution and initial melting with solution temperature. As can be seen from Fig.3, the eutectic structure gradually dissolves with the increase of solution temperature, and the eutectic structure completely dissolves when heated to 1315℃. When the solution temperature reaches 1335℃, the initial melting occurs in the alloy, and the fraction of the initial melting volume increases gradually with the increase of heating temperature. The temperature range between solution γ′ phase temperature and initial melting temperature of the alloy is usually called "heat treatment window". Solution treatment temperature should be selected in the heat treatment window to ensure that the γ′ phase can be dissolved into the γ phase without initial melting, eliminate eutectic (γ+γ′) and achieve homogenization of alloy elements. As can be seen from Fig.3, the "heat treatment window" of the alloy is 1315℃~1330℃. The lowering of initial melting temperature obtained by metallographic method by 10℃ is set as the highest solid solution temperature, which can not only improve the efficiency of solid solution heat treatment to the greatest extent, but also prevent the initial melting of alloys caused by temperature fluctuation during the heating process of heat treatment furnace. Stepped solid solution heat treatment method is adopted. By comparing the structures after different time solid solution treatment, considering the time and effect of solid solution, the solid solution heat treatment system as listed in the Table 2 is finally determined.

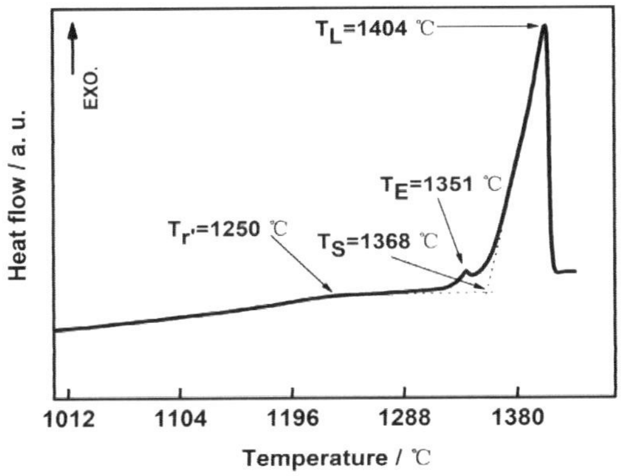

Fig.2. DSC heating results of the as-cast alloy.

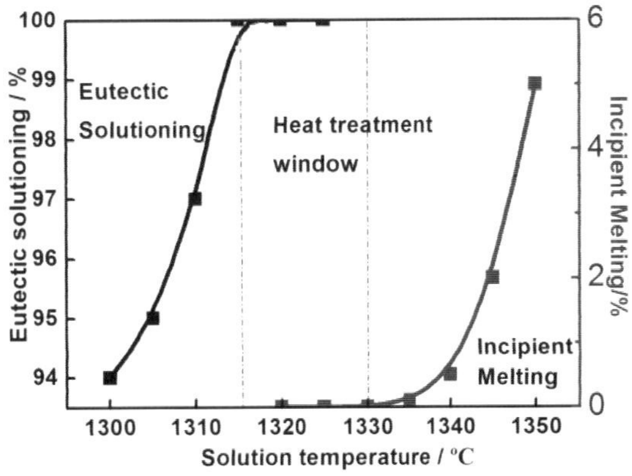

Fig.3. Changes of volume fraction of eutectic dissolution and initial melting with the increasing of solution temperature.

3.2. Microstructure after alloy solution heat treatment

Fig. 4 (a) shows the microstructure of the alloy after solution heat treatment. It can be seen that the eutectic of the γ/γ' two-phase of the alloy has been completely eliminated. But due to the high content of refractory elements such as Re, W and Ta in the alloy, dendrite cores patterns can still be observed in the solution heat treatement microstructure. Fig. 4(b) and 4(c) show the scanned tissue of the dendrite cores and interdendritic region after alloy solution heat treatment. It can be seen that the shape of the γ' phase in the dendrite cores, and inter-dendritic region after alloy solution heat treatment is more regular and average than the as-cast state and the size of the γ' phase becomes more finer. According to statistics, the average size of the dendrite cores γ' phase is about 80 nm, and the average size of the interdendrites γ' phase is about 120 nm.

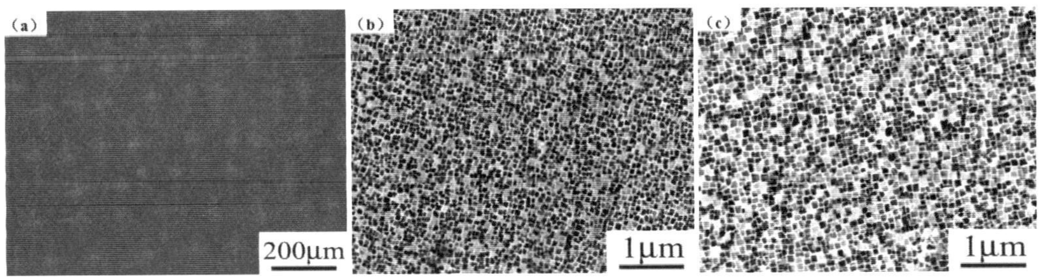

Fig. 4. Microstructures of solid solution alloy: (a)Optical microscopy image; (b) SEM image of dendrite cores; (c) SEM image of interdendrites.

Fig. 5. is a comparison of the segregation coefficients of the elements before and after the solution heat treatment of the alloy. It can be seen that the degree of segregation of all alloying elements after solution heat treatment has been reduced distinctly. The segregation type of each alloying element did not change, Re and W are still negative segregation elements, and Mo, Al and Ta are still positive segregation elements. After solution heat treatment, Re is still the most serious negative segregation element. The segregation coefficient is changed from 1.87 to 1.46; W is a more serious negative segregation element. The segregation coefficient is changed from 1.60 to 1.22; Ta is the most serious positive segregation element. The segregation coefficient is changed from 0.65 to 0.87; Al and Mo are weaker positive segregation elements, and the segregation ratio is closer to 1 after solution heat treatment.

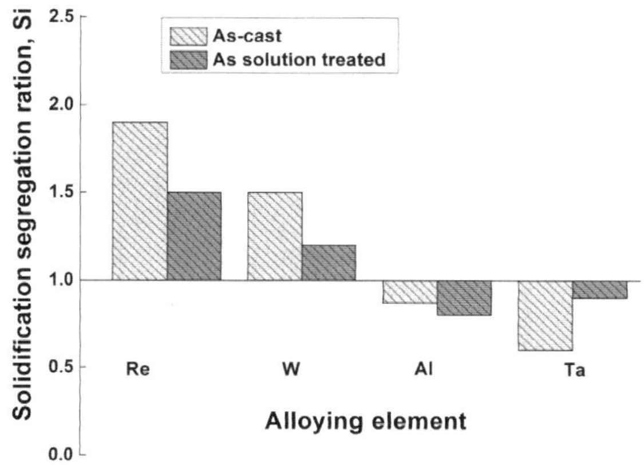

Fig. 5. Comparison of segregation of various elements before and after solution heat treatment.

Therefore, the high temperature solution heat treatment significantly improves the uniformity of the microstructure and composition of the dendrite cores and interdendritic region of the alloy. It significantly reduces the difference between the precipitation strengthening and the solid solution strengthening effect in different region of the alloy, thereby improving the dendrite cores and branches of the alloy. The continuity and uniformity of the inter granular strength play a significant role in the improvement of the mechanical properties of the alloy.

3.3. Alloy after aging heat treatment

Fig. 6. shows the typical microstructure of the alloy dendrite cores after different the first-stage aging treatment ((1100°C~1200°C)/4h, AC), and after the same secondary aging treatment (870°C/32h, AC). From the aspect of the phase of the γ' phase, when the alloy aging temperature is 1100°C, the cubicity of the γ' phase is poor. The cubicity of the γ' phase is gradually increased and the arrangement of the γ' phase is more neat when the alloy aging temperature is 1120°C and 1140°C. When the aging treatment temperature is 1160°C, the alloy cubicity begins to decrease. The sharp corners of the γ' phase begin to dissolve, making the cube smooth. The morphology of the γ' phase is similar to that of the spherical shape and the precipitation of the fine γ' phase is present in the wide γ matrix channel when the aging treatment temperature is 1180°C and 1200°C. When the aging treatment temperature is 1100°C~1140°C, the average size of the γ' phase becomes insignificant and approximately 0.32μm. However, when the aging temperature exceeds 1160°C, the average size of the γ' phase gradually increases and the effective temperature is 1200°C, the γ' phase coarsening becomes serious. The average size of the γ' phase is 0.8μm. With the increase of the first-age aging temperature, the volume fraction of the strontium between the dendrite cores and the interdendritic region increased first and then decreased, reaching a peak at 1140°C and 1160°C.

Fig. 7. shows the changes in the size and volume fraction of the dendritic core γ' phase of the alloy after different aging heat treatments. When the aging temperature is 1100 °C to 1140 °C, and the average size change of the γ' phase is not significant, about 0.32μ m. When the aging temperature exceeds 1160 °C, the average size of the γ' phase gradually increases, the effective temperature is 1200 °C, the γ' phase coarsens seriously, and the average size of the γ' phase reaches 0.8 μm. As the first-stage aging treatment temperature increases, the volume fraction of the dendritic core γ' phase increases first and then decreases, reaching a peak at 1160 °C (Fig. 7.), and the peak value is about 66%.

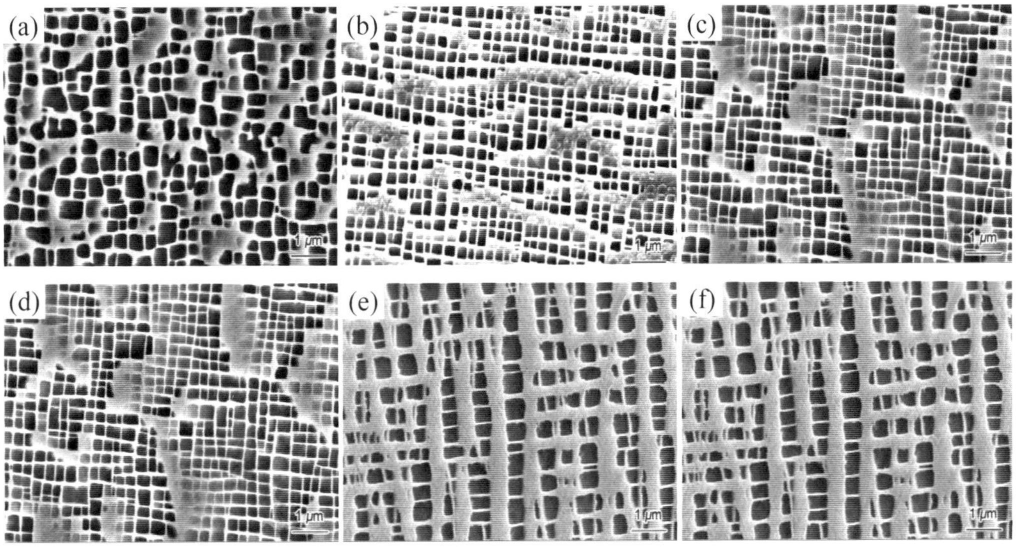

Fig. 6. Microstructure of the alloy dendrite cores after different the first-stage aging treatment: (a) 1100℃, (b) 1120℃, (c) 1140℃, (d) 1160℃, (e) 1180℃ and (f) 1200℃.

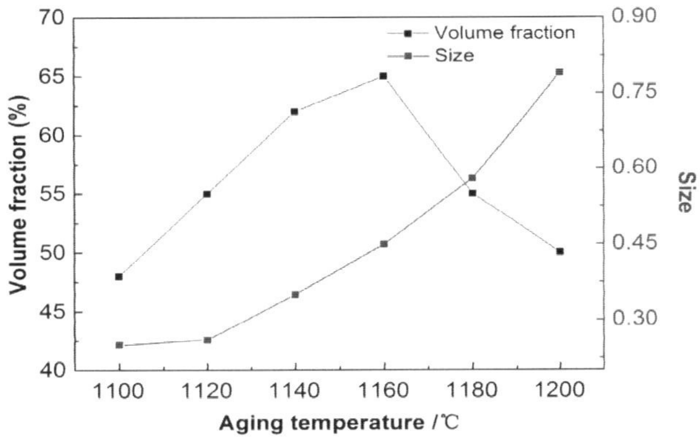

Fig. 7. Changes in size and volume fraction of the dendritic core γ' phase of the alloy after different aging heat treatments.

4. Discussion

The single-phase solid solution is firstly precipitated from the liquid phase when the nickel-based single crystal superalloy is solidified. The single-phase solid solution is dendritic growth and the high melting point elements such as Re and W are enriched to in dendrite cores At the same time, elements such as Al and Ta are concentrated in the liquid phase between the interdendritic region. As the liquid phase between the interdendrites solidifies, the concentration of the solute in a small amount of the remaining liquid phase reaches the eutectic point, forming the γ/γ' eutectic. The content of as-cast eutectic can reflect the degree of solidification segregation [1]. In this work, about 6% of the γ/γ' eutectic exists in the as-cast microstructure (Fig. 1.), and as-cast segregation is also severe, and the solidification segregation coefficient of Re reaches 1.87 (Fig. 5.).

Pretreatment of as-cast alloy by 1290°C/2h+ 1300°C/2h+1310°C/2h can reduce the degree of segregation of low melting point elements in the interdendritic region, and avoid the initial melting of the alloy. Subsequently, the high temperature solution treatment at 1320°C/6h is carried out, which makes the γ/γ' eutectic structure and the coarse γ' phase in the as-cast alloy completely dissolve into the γ matrix. Meanwhile, the insoluble alloying elements are sufficiently diffused at a high temperature which reduces the solidification segregation of the alloy and increases the degree of homogenization of the alloy composition (Fig. 5.). In the subsequent solid solution air cooling process, a large number of fine nanoscale γ' phase particles are precipitated from the supersaturated γ matrix. That is because the elements such as Ta and Al in the interdendritic region are still higher than the dendrite cores. Ta and Al are the main forming elements of the γ′ phase and size of the γ′ phase between the interdendritic region after solution heat treatment is significantly larger than that of the dendrite cores (Fig 4(b) and (c)). Wang et al. [13] believed that the fine γ' phase precipitated after solid solution maintained a coherent interface with the γ matrix. At the same time, since the cubic γ' phase precipitated is small in size and large in number and there are many interfaces of γ'/γ two phases, so there is a large interfacial energy and a certain degree of lattice strain energy in the alloy.

During the first-stage aging treatment of the alloy, the super-saturated γ' phase particles in the alloy continue to grow under the joint action of the coherent interface strain energy. During the aging treatment period from 1100°C to 1140°C, the morphology of the γ' phase is limited by the interfacial energy and coherent interface strain energy between the two phases. The γ' phase does not grow excessively. The γ/γ' phase maintains a coherent interface. The interaction of the interfacial energy and the coherent interface strain energy causes the γ' phase to grow into a cubic shape, as shown in (Fig.6(a)~(c)). When the temperature is improved, the γ'/γ phase is no longer flat, which indicates that the γ'/γ two-phase coherent interface disappears and the strain energy is weakened. Therefore, under the action of the interface energy, the γ' phase can be promoted to grow into a spherical or thick strip-like morphology (Fig. 6(e)~(f)).

5. Conclusion

The alloy in as-cast is exhibited a typical dendrite structure and there are approximately 6% the γ/γ′ eutectic between the interdendritic region. The gold liquidus and solidus temperatures of the alloy are 1404°C and 1368°C respectively. The γ' phase solid solution temperature is 1250°C.

The range of solid solution heat treatment of the alloy is 1315℃~1330℃. After 1320℃/6h of solid solution heat treatment, the segregation of the alloy elements becomes alleviated. The γ/γ′ eutectic is basically eliminated and the fine γ' phase is precipitated in the solid solution and air cooling process.

As the first-stage aging treatment temperature increases, the size of the γ' phase gradually increases. The volume fraction of the γ' phase first increases and then decreases. The cubicity of the γ' phase decreases when the aging treatment temperature reaches or exceeds 1160°C. The γ matrix channel becomes significantly coarsened and the fine γ' phase is precipitated in the channel when the first-stage treatment aging temperature are 1180°C and 1200°C.

Acknowledgements

This work was supported by from the National Natural Science Foundation of China (Grant No.51605386) and the AECC Sichuan Gas Turbine Establishment. These supports are gratefully acknowledged.

References

[1] Y S Zhao, J Zhang, Y S Luo, et al. Effect of Hf on high temperature low stress rupture properties of a second generation Ni-based single crystal superalloy DD11[J]. Acta Metall Sin, 2015(10): 1261-1272.
[2] W S Waltson, K S O'Hara, E W Ross, et al. René N6: Third generation single crystal superalloy. Superalloy 1996. Warrendale, PA, 1996: 27-34.
[3] W S Waltson, Schaeffer, J C, W H Murphy. A new type of microstructural instability in superalloy - SRZ. Superalloy 1996. Warrendale, 1996: 9-18.
[4] Y H Kong, Q Z Chen, Q Z Knowles. Effects of minor additions on microstructure and creep performance of RR2086 SX superalloy. Journal of Materials Science, 2004, 39(23): 6993-7001.

[5] G Liu, L Liu, S X Zhang, et al. Effects of Re and Ru on microstructure and segregation of Ni-based single-crystal superalloy[J]. Acta Metall Sin, 2012, 48(7): 845-852.
[6] B C Wilson, G E Fuchs. The effect of composition, misfit, and heat treatment on the primary creep behavior of single crystal nickel base superalloy PWA 1480 and PWA 1484. Warrendale: Minerals, Metals & Materials Soc, 2008:
[7] P Caron, T Khan, Y Ohta, et al. Creep deformation anisotropy in single crystal superalloy. Superalloy 1988, 1988: 215-224.
[8] P K Caron. Improvement of creep strength in a nickel-base single-crystal superalloy by heat treatment. Materials Science and Engineering A, 1983, 61(2): 173-184.
[9] R Mackay, L Ebert. The development of γ-γ' lamellar structures in a nickel-base superalloy during elevated temperature mechanical testing. Metallurgical and Materials Transactions A, 1985, 16(11): 1969-1982.
[10] T Murakumo, T Kobayashy, Y Koizumi, et al. Creep behaviour of Ni-base single-crystal superalloy with various γ' volume fraction. Acta Materialia, 2004, 52(12): 3737-3744.
[11] G E Fuchs. Improvement of creep strength of a third generation, single-crystal Ni-base superalloy by solution heat treatment. Journal of Materials Engineering and Performance, 2002, 11(1): 19-25.
[12] K Harris, J B Wahl. Improved single crystal superalloy, CMSX-4 (R)(SLS)[La+Y] and CMSX-486 (R). In: Green KA, Pollock TM, Harada H, Howson TE, Reed RC, Schirra JJ, et al. (eds). Superalloy 2004. Warrendale: Minerals, Metals & Materials Soc, 2004: 45-52.
[13] M G Wang, S G Tian, X F Yu, et al. Effects of heat treatment on composition segregation and enduring properties of single crystal nickel-based superalloy[J]. Journal of Shenyang University of Technology, 2009, 31(6): 525-530.

Constitutive model and numerical simulation of HSLA steel during isothermal compression

Ning Li [a,b], **Yao Huang** [a,b], **Hui Zhang** [a,b], **Renheng Han** [a,b], **Chengzhi Zhao** [a,b,*], **and Hexin Zhang** [a,b,*]

[a] College of Materials Science and Chemical Engineering, Harbin Engineering University, Harbin 150001, China
[b] Key Laboratory of Superlight Materials and Surface Technology, Ministry of Education

E-mail: zhaochengzhi@hrbeu.edu.cn, zhanghx@hrbeu.edu.cn

Abstract. The true stress-strain curve of HSLA steel was obtained though a MMS-200 experiment machine. The flow behavior of HSLA steel during isothermal compression at different deformation conditions was investigated. The result indicated the flow stress increases as the strain rate increases or the deformation temperature decreases, and the flow curves show obvious dynamic softening characteristics at low strain rate or high temperature. The peak stress constitutive equation was established and the activation energy be calculated as 437955 J/mol. The effective strain and flow stress during isothermal compression were studied by finite element method. The simulation results show that the effective strain distribution at different regions is inhomogeneous. The flow stresses obtained by FEM are consistent with the corresponding test results.

1. Introduction

HSLA (high strength low alloy) steel has interested great attention from researchers owning to its excellent corrosion resistance and mechanical properties, and is used as structural components widely in automobiles, bridges, ships, pipelines, pressure vessels, and containers in recent years [1, 2]. Steel with low carbon content and trace addition of strong carbides, such as V, Nb, and Ti, is called microalloyed steel [3, 4]. The grain of microalloyed steel can be significantly improved by TMCP (thermo mechanical control process) technology. HSLA steels' mechanical properties can be enhanced using combining microalloying with TMCP technology [5, 6]. It thus is very significant to study HSLA steel's thermal deformation behavior for improving mechanical properties. In general, most researchers used a constitutive model based on the peak stress or constitutive model compensated with strain to study materials' thermal deformation behavior. The constitutive model that is based on peak stress can describe peak stress' fluctuation only with temperatures and strain rates, and neglect strain's effect on stress. Although constitutive equations compensated with strain can be used to study the relationship among deformation temperatures, flow stress, strains and strain rates, the process of establishing the equation is too complicated and the calculation results of flow stress at a large strain rate are not very accurate. With computer technology's development, the numerical simulation has become an effective method to predict the thermal deformation behavior during isothermal compression [7, 8]. However, it is rarely reported that the flow behavior of HSLA steel during hot compression is studied by numerical simulation.

In this paper, HSLA steel's true stress-strain curves were obtained by a MMS-200 experiment machine. The flow behavior of HSLA steel during isothermal compression at different deformation conditions was investigated. The material constants were calculated and the constitutive model was established. The numerical simulation was used to predict the effective strain distribution and flow stress curves of HSLA steel during the isothermal compression process.

2. Experimental procedures

The investigated steel is HSLA steel, the composition (wt%) of this steel used was as follows: 0.013 C, 0.147 Si, 0.203 Cr, 0.534 Ni, 0.392 Cu, 0.135 Mo, 0.041 Nb, 1.17 Mn, and balance Fe. The size of the cylindrical specimen is $\Phi 8 * 15$ mm, and the specimens were cut by wire cut machine. The hot compression tests were performed by a MMS-200 test machine at temperatures of 900, 1000 and 1100 °C, and strain rates of 0.01-1 s^{-1}. Fig. 1 is a schematic diagram for isothermal compression. Specimens were heated to 1250 °C at 20 °C/s and holding for 180 s to obtain a uniform austenite structure. Then the temperature was reduced to the isothermal compression temperature at 10 °C/s and the temperature gradient was eliminated by holding for 20 s. When the true strain is compressed to 0.7, all specimens are quenched immediately. The deformed specimen was cut along the compression axis. After grinding and polishing, the axis section was etched in ethanol solution containing 5% nitric acid.

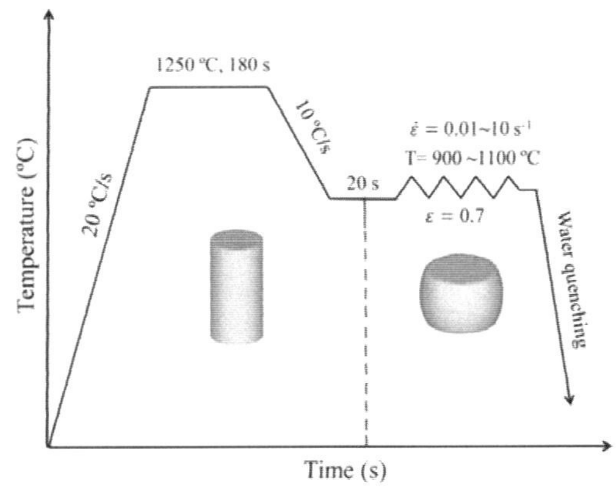

Fig. 1. The schematic diagram for hot compression test.

3. Results and Discussions

3.1. *Hot deformation behaviour*

The flow curves of the steel at varying strain rates and temperatures are presented in Fig. 2. It is observable that deformation temperatures and strain rates have important effects on flow stress. The flow stress increases as the strain rate increases or the deformation temperature decreases. At a definite strain rate, the flow stress increases with decreasing temperature. Dynamic recovery and DRX were performed more fully at a higher temperature, resulting in more obvious dynamic softening effect. Under a given temperature, with increasing strain rate, the flow stress was reduced. As demonstrated in Fig. 2 (d), at higher strain rates, dynamic recovery and dynamic recrystallization have no enough time to fully perform due to short deformation time. Therefore, compared to low strain rate, the dynamic softening effect is not obvious at the high strain rates.

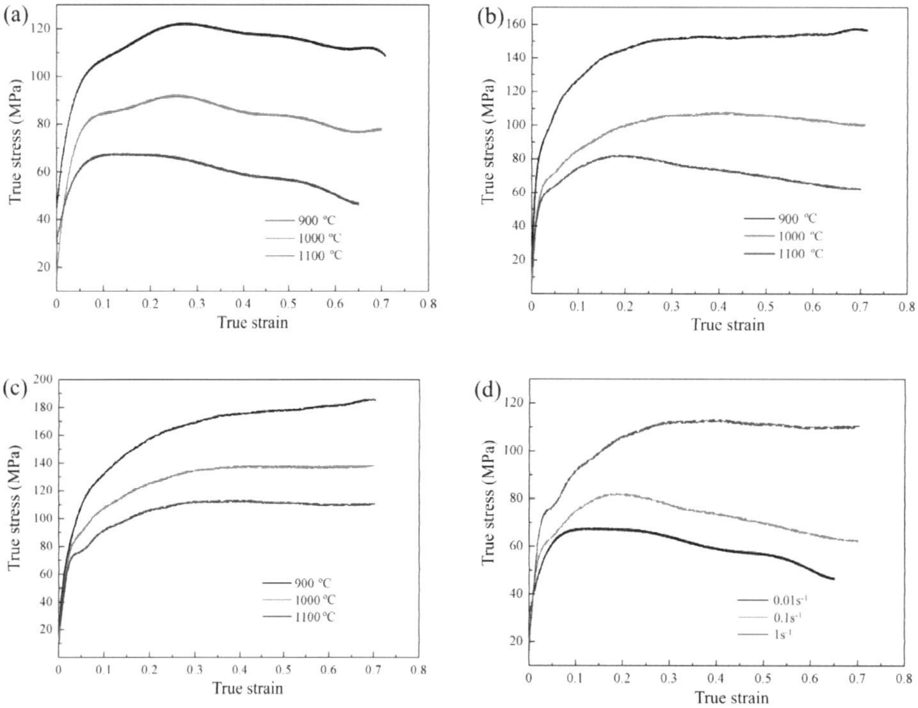

Fig. 2. The flow curves of HSLA steel at varying strain rates and temperatures: (a) 0.01s^{-1}, (b) 0.1 s^{-1}, (c) 1 s^{-1}, (d) 1100 °C.

3.2. Constitutive equation

The flow stress is primarily influenced by the temperature, strain and strain rate, and the temperature and strain rate play a major role. The relationship between deformation temperature, flow stress and strain rate can be presented by the following kinetic equation.

$$Z = \dot{\varepsilon}\exp(Q/RT) \tag{1}$$

$$\dot{\varepsilon}\exp(Q/RT) = A_1{\sigma_P}^{n_1} \quad (when\ \alpha\sigma < 0.8) \tag{2}$$

$$\dot{\varepsilon}\exp(Q/RT) = A_2\exp(\beta\sigma_P) \quad (when\ \alpha\sigma > 1.2) \tag{3}$$

$$\dot{\varepsilon}\exp(Q/RT) = A[\sinh(\alpha\sigma_P)]^n \quad (for\ all\ \sigma) \tag{4}$$

In which, Z is the Z parameter, T is the hot deformation temperature (K), $\dot{\varepsilon}$ is the strain rate (s^{-1}), σ_P is the peak stress (MPa), Q is apparent activation energy ($J\cdot mol^{-1}$), R is the ideal gas constant (8.314 $J\cdot mol^{-1}\cdot K^{-1}$), n_1, β, A_1, A_2, A, and n are the material constants, where $\alpha = \beta/n_1$.

Taking the logarithm of the two sides of the Eqs. (2) and (3), and Eqs. (2) and (3) can be rewritten to Eqs. (5) and (6), respectively.

$$\ln\dot{\varepsilon} = n_1\ln\sigma_p + \ln A_1 - Q/RT \tag{5}$$

$$\ln\dot{\varepsilon} = \beta\sigma_p + \ln A_2 - Q/RT \tag{6}$$

The linear relation of $\ln\dot{\varepsilon} - \ln\sigma_p$ and $\ln\dot{\varepsilon} - \sigma_p$ at different strain rates are fitted, the fitting results are shown in Fig. 3a and 3b. The average slopes are used to calculate n_1 and β, which are 10.680 and

0.0927. According to $\alpha = \beta/n_1$ =0.00868, the value of α can be calculated. When the deformation temperature is constant, Eq. (7) can be obtained by partial differential of Eq. (4).

$$n = \left[\frac{\partial \ln \dot{\varepsilon}}{\partial \ln[\sinh(\alpha\sigma_p)]}\right]_T \tag{7}$$

Similarly, at a given strain rate, Eq. (8) can be obtained by partial differential of Eq. (4).

$$Q = nR\left[\frac{\partial \ln[\sinh(\alpha\sigma_p)]}{\partial(1/T)}\right]_{\dot{\varepsilon}} \tag{8}$$

$$S = \left[\frac{\partial \ln[\sinh(\alpha\sigma_p)]}{\partial(1/T)}\right]_{\dot{\varepsilon}} \tag{9}$$

The values of n and S are obtained through the linear fitting slopes of Fig. 3c and 3d. The average values of n and S are 7.758 and 6047, respectively. Accordingly $Q = nRS$, the value of activation energy is calculated as 437955 $J \cdot mol^{-1}$. By combining Eq. (4) and Eq. (1), Eq. (10) can be obtained. Taking the logarithm of the two sides of Eq. (10), and Eq. (10) is rewritten to Eq. (11).

$$Z = A[\sinh(\alpha\sigma_P)]^n \tag{10}$$

$$\ln Z = \ln A + n\ln[\sinh(\alpha\sigma_p)] \tag{11}$$

Fig. 4 shows a good linear relationship between $\ln Z$ and $\ln[\sinh(\alpha\sigma_P)]$. $\ln A$ is the intercept of the fitted curve, and $\ln A$ is calculated as 37.98 by linear fitting. The calculated results of material constants and Q are given in Table 1. Therefore, the constitutive equations of the steel based on peak stress can be obtained as following:

$$\dot{\varepsilon} = 3.12 \times 10^6[\sinh(0.00868\sigma_p)]^{7.758}\exp\left(-\frac{437955}{RT}\right) \tag{12}$$

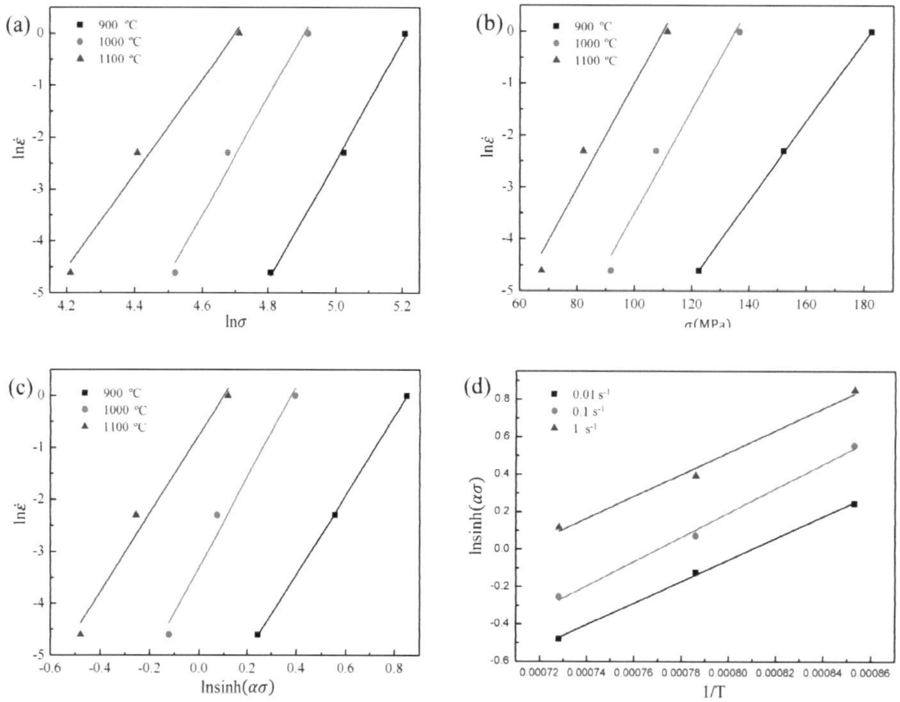

Fig. 3. Material constants solution for HSLA steel: (a) $\ln\dot{\varepsilon} - \ln\sigma$, (b) $\ln\dot{\varepsilon} - \sigma$,(c) $\ln\dot{\varepsilon} - \ln[\sinh(\alpha\sigma_p)]$, (d) $\ln[\sinh(\alpha\sigma_p)] - 1/T$.

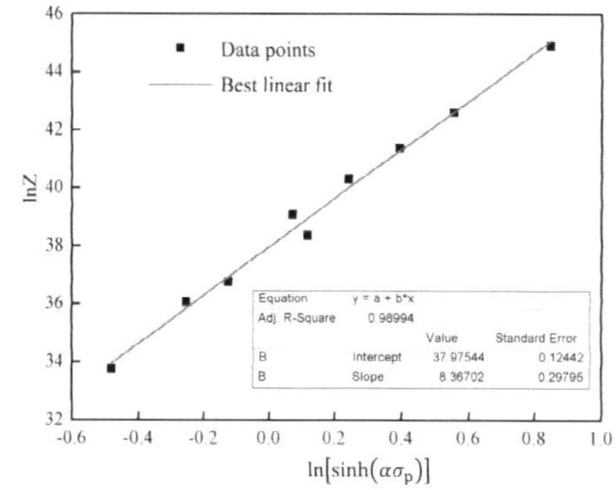

Fig. 4. Linear relationship between lnZ and $\ln[\sinh(\alpha\sigma_p)]$.

Table 1 Material constants for HSLA steel.

Parameter	n_1	β	α (Mpa^{-1})	n	Q ($J\cdot mol^{-1}$)	lnA
Value	10.680	0.0927	0.00868	7.758	437955	37.98

3.3. FEM simulation

Isothermal compression process for HSLA steel was simulated by using FEM (finite element method) under different deformation conditions. The flow stress data at varying deformation condition was obtained by isothermal compression test. The geometric model for FEM simulation is shown in Fig. 5.

The workpiece and dies were set as plastic type and rigid type, respectively. The friction type between workpiece and dies was set to shear type, the coefficient of friction is set to 0.6 [9]. The physical parameters for finite element analysis were calculated by Jmatpro software, and the calculated results are shown in Fig. 6. The commercial software package Deform-3D is used to simulate isothermal compression.

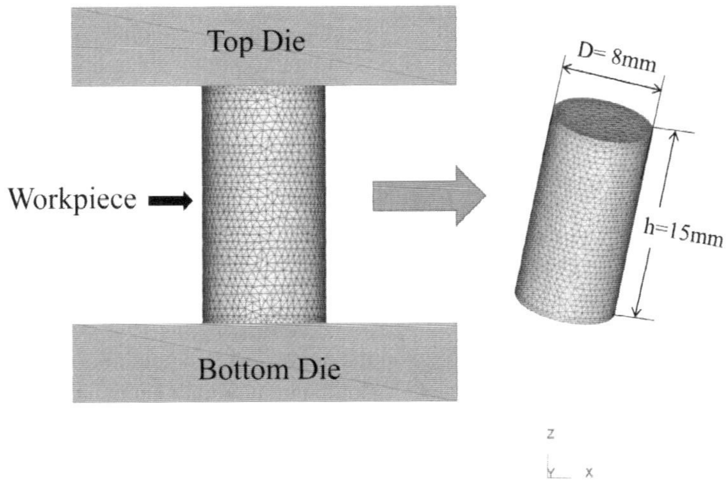

Fig. 5. FEM model of isothermal compression process.

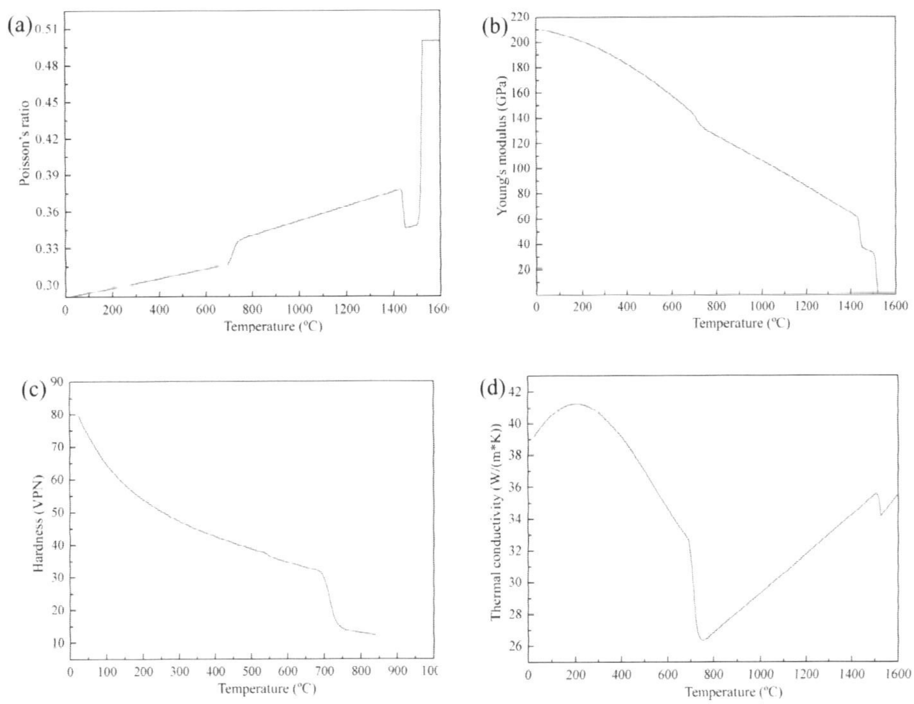

Fig. 6. Relationship between thermo-physical parameters and temperature of HSLA steel.

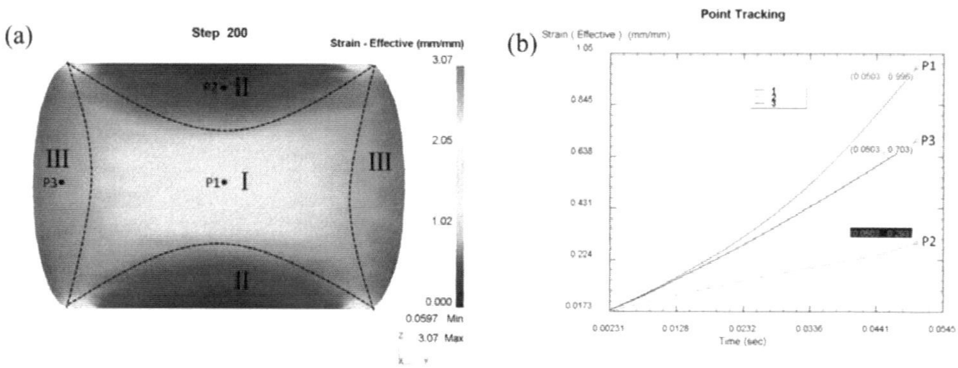

Fig. 7. (a) Effective strain distribution at different regions, (b) the effective strain-time curves of the Point 1, Point 2 and Point 3.

The strain distributions of deformed specimen have a major impact on the material's microstructure. Therefore, it's necessary to study deformed specimen's internal strain distribution. The effective strain distribution on the deformed specimen's axial section is shown by Fig. It shows that the effective strain is unevenly distributed. According to the effective strain distribution, the axial section is divided into three regions [10]: (1) Domain I, which was called the large deformation region with relatively large level of deformation at the center of specimen; (2) Domain II, which was the bonded region at both ends of the deformed specimen, this region has relatively small deformation due to friction between the workpiece and the die; (3) Domain III was called free deformation region, which is located at the edge of the deformed specimen. The deformation level in domain III is higher than that in domain II, but lower than that in domain I. To further study the changes of effective strain under different regions during the hot deformation process, Point 1, Point 2 and Point 3 were selected as the feature points in domain I, domain II and domain III. Fig. 7 (b) is the effective strain vs. time curves of the selected points. It indicated that the effective strain at point 1 increases rapidly with time, and the maximum effective t strain is 0.996. The effective strain at point 2 increases slowly with time, and the maximum effective strain is only 0.293. The effective strain at point 3 during deformation is higher than that at point 2 but lower than that at point 1, and the maximum value of the effective strain is 0.703. Fig. 8 shown the simulation results of stress-time curve at varying deformation conditions. It indicates the effective stress increases rapidly with time in initial stage. When the peak stress is reached, the effective stress gradually stabilizes with time. The variation trend of flow stress is consistent with the test data. To further verify the accuracy of simulation results. According to Eq. 13, the simulated stress-time curves can be converted into the stress-strain curves, and compared with the experimentally obtained flow curves. The comparison between experimental results and stress-strain curve simulation results at varying deformation conditions are shown in Fig. 9. The results indicate that the finite element analysis results are consistent with the corresponding test results.

$$\varepsilon = -\ln(1 - \dot{\varepsilon}t) \tag{13}$$

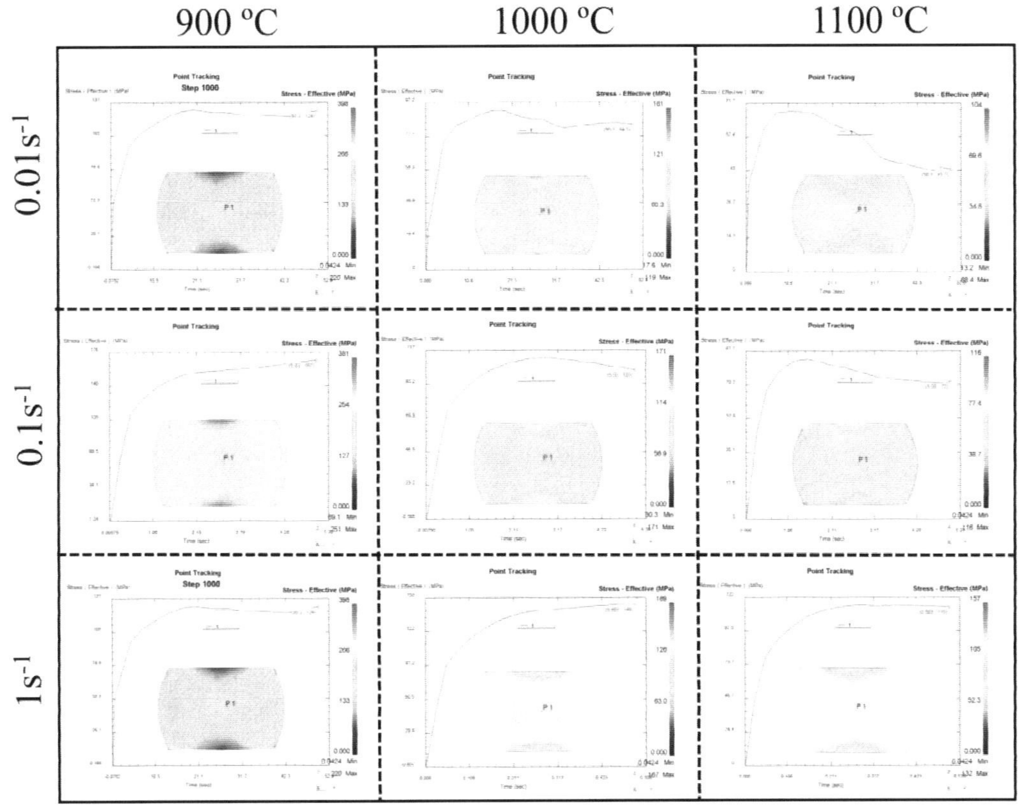

Fig. 8. The simulation results of the stress-time curves at varying strain rates and deformation temperatures.

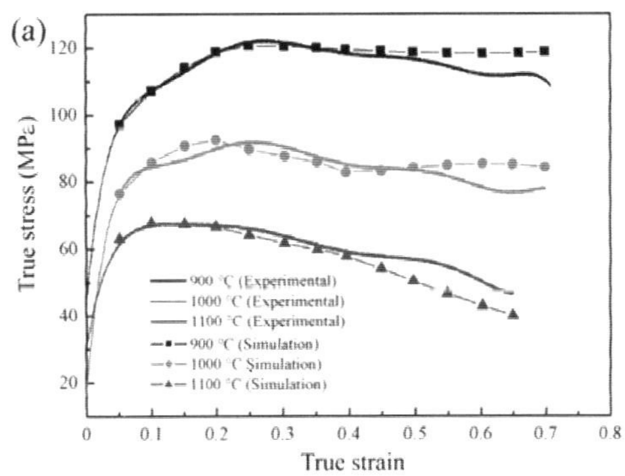

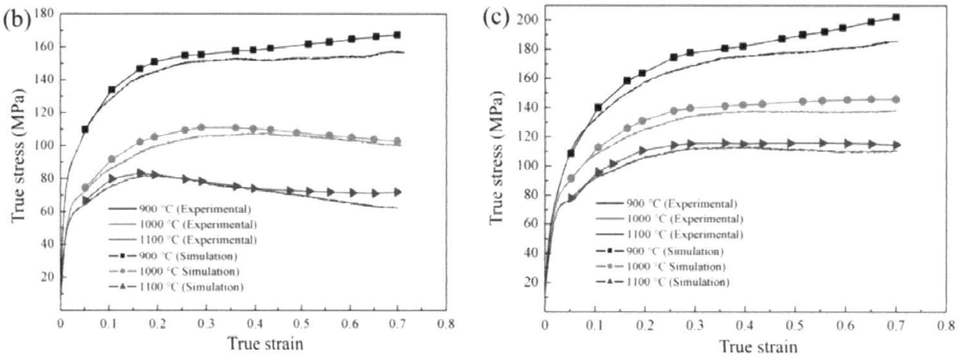

Fig. 9. Comparison between the simulation and experimental true stress-strain curves at different temperatures and strain rates: (a) 0.01 s^{-1}, (b) 0.1 s^{-1}, (c) 1 s^{-1}.

4. Conclusions

In this paper, the flow behavior of HSLA steel during isothermal compression at different deformation conditions was investigated, the flow stress increases as the strain rate increases or the deformation temperature decreases. The peak stress constitutive equation was established and the apparent activation energy be calculated as 437955 J/mol. The effective strain distribution and flow stress of HSLA steel during isothermal compression were investigated by using FEA method, and the stress-strain curves obtained by FEM are consistent with the corresponding test results.

Acknowledgements

This paper is funded by the International Exchange Program of Harbin Engineering University for Innovation-oriented Talents Cultivation.

References

[1] X. Chen, Y. Huang, J. Alloy. Compd. 619 (2015) 564-571.
[2] R. Shukla, S.K. Das, B.R. Kumar, S.K. Ghosh, S. Kundu, S. Chatterjee, Metall. Mater. Trans. A-Phys. Metall. Mater. Sci. 43A (2012) 4835-4845.
[3] A. Galibois, M.R. Krishnadev, A. Dubé, Metall. Trans. A 10 (1979) 985-995.
[4] A. Momeni, H. Arabi, A. Rezaei, H. Badri, S.M. Abbasi, Mater. Sci. Eng. A 528 (2011) 2158-2163.
[5] M. Klein, H. Spindler, A. Luger, R. Rauch, P. Stiaszny, M. Eigelsberger, Mater. Sci. Forum. 500-501 (2005) 543-550.
[6] H.L. Wei, G.Q. Liu, X. Xiao, H.T. Zhao, H. Ding, R.M. Kang, Mater. Sci. Eng. A 564 (2013) 140-146.
[7] M.R. Rahul, S. Samal, S. Venugopal, G. Phanikumar, J. Alloy. Compd. 749 (2018) 1115-1127.
[8] L. Zhou, C. Cui, Q.Z. Wang, C. Li, B.L. Xiao, Z.Y. Ma, J. Mater. Sci. Technol. 34 (2018) 1730-1738.
[9] Z.J. Wang, L.H. Qi, G. Wang, H.J. Li, M.S. Dargusch, Mech. Mater. 102 (2016) 90-96.
[10] Y. Xu, C. Chen, X.X. Zhang, H.H. Dai, J.B. Jia, Z.H. Bai, Mater. Charact. 145 (2018) 39-52.

IOP Conf. Series: Materials Science and Engineering **605** (2019) 012019 doi:10.1088/1757-899X/605/1/012019

The mechanical oscillation of a single carbon nanocoil driven by a focused laser beam

Y L Liu[1*], J Shen[2*], and Y M Sun[3]

[1]Dalian University of Science and Technology, No. 999-26, Hong Kong Road, Lvshun Economic Development Zone, Dalian, Liaoning Province, 116052, China

[2]Dalian Neusoft University of Information, No.8 Software Park Road, Dalian 116023, China

[3]College of Electronic Science and Technology, Shenzhen University, Nanhai Ave 3688, Shenzhen, Guangdong, 518060, China

* Corresponding author's e-mail: shenjian@neusoft.edu.cn; liuyuli2005@163.com

Abstract. We have determined the mechanical properties of a long single carbon nanocoil (ALS-CNC) interacting with a focused laser beam. The mechanical oscillation properties of ALS-CNC in liquid based on the interaction of a CNC with light have been demonstrated. Considering the viscous force of a liquid, ALS-CNC oscillation does not occur by laser irradiation when ALS-CNC is in air. A CNC switch controlled by laser irradiation was realized. The first section in your paper.

1. Introduction

Research on light-matter interactions has focused on its applications in various fields, such as biochemistry, sensing and clinical medicine [1-9].

Optical trapping of nano/micro-particles has been researched for many years in some fields such as biology [13,14]. Recently, photophoretic force, which is called light-driven pulling force, due to thermal effect, has been brought into focus about the light-matter interaction research. The resultant forces of them is discussed to manipulate metal objects [10-12]. Photophoretic forces or optical gradient forces are considered as the main reason of objects moving in most cases, as doing in different works [21,22,23]. However, few works research the resultant force of these two kinds of forces for explaining experiment result.

Carbon nanocoils with a unique helical morphology had excellent physical properties and were expected to be applied as living molecular components and nano-spring etc. [15,16,17,18,19,20].

Recently, carbon nanocoil interactions with light is of significant interest which can realize many novel technologies.

Our previous work showed that micro-bubbles can be generated by laser irradiation on carbon nanocoils thoroughly immersed in water [27] and paraffin microspheres can be driven on carbon nanocoils in air by a focused laser beam [28], which is the result of the conversion of light energy to heat energy. In addition, Carbon nanocoils manipulated by a focused laser beam has been reported by our research team [29,30]. However, the light used in those technologies must be focused precisely on the CNC body. The laser focal point should not deviate from the CNC body.

In 1969, Ashkin reported that an optical gradient force that is parallel to the direction of the laser beam can be applied to micro-particulates near a focused laser beam.

The index of refraction of an object is higher than that of the surrounding medium. Optical pulling forces is applied to the object. In contrast, the index of refraction of an object is lower than that of the surrounding medium. Optical pushing forces is applied to the object. Chen et al. measured the spring constant of individual CNCs and obtained a value of 0.12 N/m, suggesting that the laser focal point deviated from the CNC body [31,32]. The CNC will be acted upon by an optical gradient force, which provides a good opportunity to research the mechanical properties of a long single carbon nanocoil (ALS-CNC) interacting with a focused laser beam.

In this article, we present the mechanical oscillation properties of ALS-CNC driven by focused laser beam. We also show that the potential applications of the light- CNC interaction which may give some interest phenomenon which is attractive.

2. Experimental details

Carbon nanocoils of average coil diameter from 300 to 600 nm were synthesized by chemical vapour deposition in this experiment[33]. In this experiment, a suspension liquid of Al_2O_3 powders was selected as as support materials and spin coated on SiO_2substrate. $Fe_2(SO_4)_3/SnCl_2$solution was used as catalyst in this experiment. Then the catalyst solutions were dropped on these substrates.Then the samples was dried at $40-50$°Cfor $10min$ and calcined at 700°C for $30min$ to oxidize the catalysts. Acetylene gas was introduced at $60sccm$ to a thermal CVD system at 700°C for $30min$ in ambient Ar gas. Finally, CNC aggregation can be acquired as shown in fig. 1.

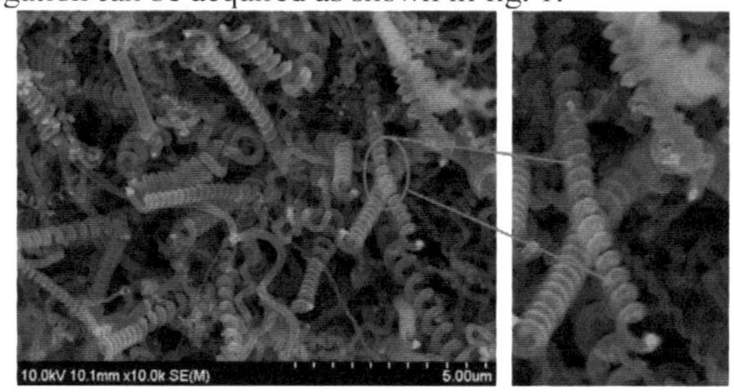

Figure 1. (a) SEM of the carbon nanocoils synthesized in this study; (b) The enlarged SEM image of a single carbon nanocoil from CNCs aggregation.

ALS-CNC with a length of 50um obtained from CNC aggregation is attached to the tip of a copper wire on slide glass [34]. Several drops of deionized water were dropped onto the tip of the copper wire on the slide glass, and thus ALS-CNC could be immersed thoroughly in water at room temperature, as schematically shown in fig. 2. The liquid condition for ALS-CNC was easily achieved due to the wettability and hydrophilicity of the slide glass.

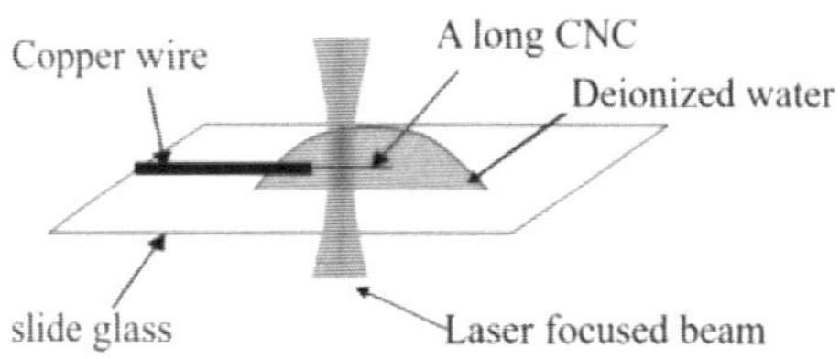

IOP Conf. Series: Materials Science and Engineering **605** (2019) 012019 doi:10.1088/1757-899X/605/1/012019

Figure 2. Diagram of the fabrication process of a long single carbon nanocoil thoroughly immersed in water.

ALS-CNC with a length of 50um is long enough that it cannot be observed inextenso from the microscopic images of a long carbon nanocoil immersed in water, as shown by the two pictures linked together in fig. 3.

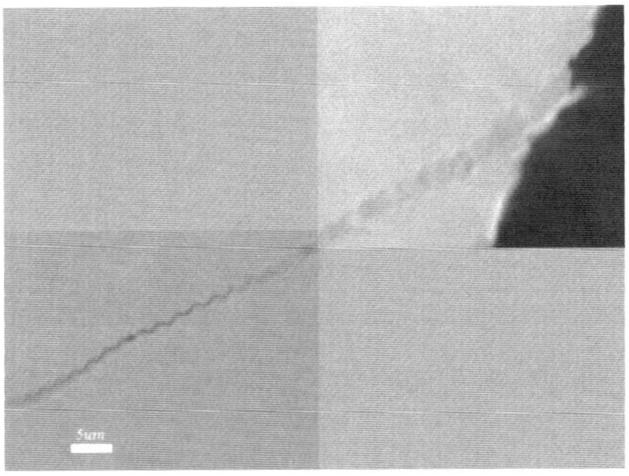

Figure 3. Microscopic image of ALS-CNC immersed in water thoroughly.

Single-beam optical tweezers system was applied to work in this study, which was described in detail in our previous work [27]. By controlling the focal distance, the laser beam spot with a 4 um diameter will be generated, which is called an optical trap. An optical trap force between the micro-object and the optical trap exists when a micro-object is in the optical trap. Charge coupled device (CCD) camera system can be used to record the process of a micro-object moving.

3. Results and discussion

With increasing distance from the tip of the copper wire on the slide glass, a clearer mechanical effect of ALS-CNC will be shown when a force is exerted on this CNC. Thus, a laser beam was focused on the position near the tip of ALS-CNC, uninterruptedly, as shown in fig. 4. The red spot (see fig. 4) images the location of the energy centre of laser light.

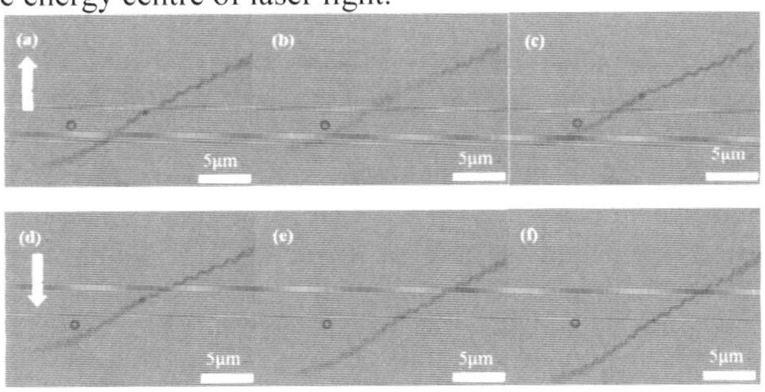

Figure 4. The process of ALS-CNC immersed in water with self-oscillations recorded by a CCD system. The red spots indicate the focal point of a laser. Laser power illumination is 4.5uW. (a) to (c) show the upper swing of a CNC; (d) to (f) show the lower swing of a CNC.

Fig. 4 (a) to (f) shows the process of ALS-CNC self-oscillations in liquid under prolonged irradiation on a position near ALS-CNC (the laser power illumination is 4.5uW in this experiment). The pictures from (a) to (f) in fig. 4 are a vibration cycle recorded by a CCD camera. The speed of vibration is very quick, and the time of a vibration cycle is less than one second. The schematic diagram is displayed in fig. 5. Therefore, no micro-bubbles can be produced at the site of laser focus due to the conversion of light energy to heat energy, which is different from the previous experiment result has been report by our team [27]. fig. 4 (a) describes the original position of the CNC untreated with laser irradiation. A laser with an illumination power of 4.5uW is focused on the red spot area near the CNC in water. In addition, the CNC self-oscillations dies down when the laser energy is zero or the site of laser focus is far from the CNC. The amplitude of the CNC self-oscillations becomes very large when the laser irradiation power increases, which is shown in fig. 6.

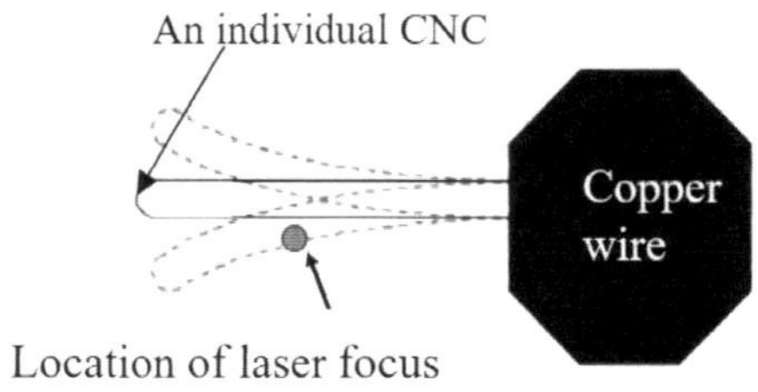

Figure 5. Schematic of a self-oscillations of ALS-CNC immersed in water induced by the focused laser.

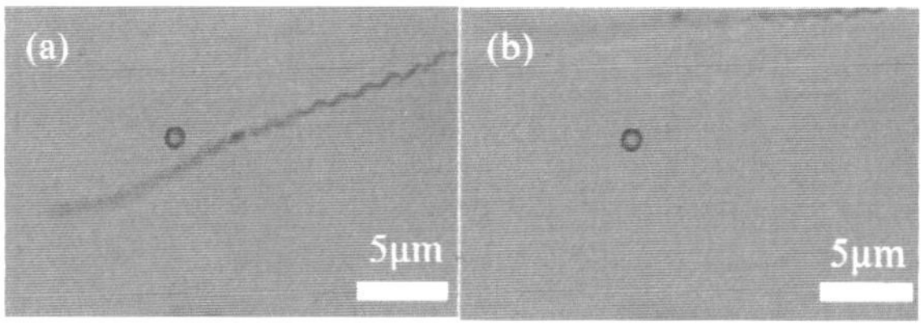

Figure 6. Effect of laser power on amplitude of CNC vibration. The red spots indicate the position of laser focus. The laser power illuminates on the CNC are 4.5uW (a) and 10.5uW (b).

Since photophoretic forces and the trapping forces of focused laser beam are perpendicular to the laser beam [31], the resultant forces of them cause ALS-CNC has to move close to or far away from the location of laser irradiation. However, CNCs exhibit outstanding flexibility due to their unique helical morphologies. Therefore, there are three forces exerted on the single CNC thoroughly immersed in water, including the trapping force from laser irradiation, the restoration force from the CNC flexibility and the damping force from the liquid (supposing the photophoretic force is ignored [20]). Hence, a long single carbon nanocoil self-oscillation occurs in water by laser irradiation.

Interestingly, laser irradiation does not cause ALS-CNC to oscillate around the end of the CNC in air. From fig. 7, ALS-CNC in air is attracted to the position of laser irradiation, and then, the CNC immediately deviates from the original location of laser irradiation under prolonged laser irradiation, as shown in fig. 7(b). Then, the CNC in the air returns to the original position when the laser energy is zero or the site of laser focus is far away from the CNC. The self-oscillation effect does not occur by laser irradiation when ALS-CNC is in air because the damping force in air is less than that in liquid, causing a difference compared to the CNC immersed in water.

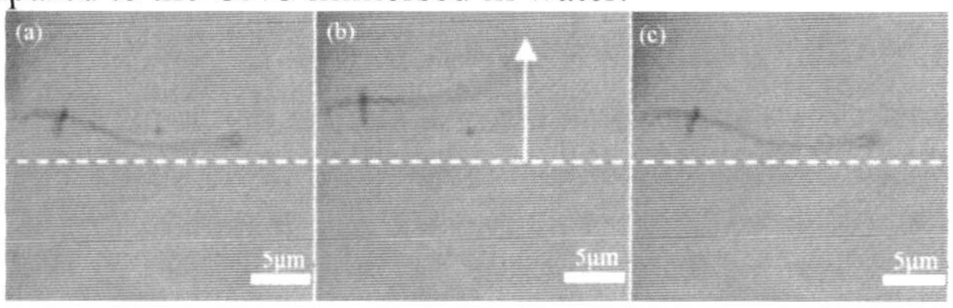

Figure 7. Microscopic images of ALS-CNC treated with laser irradiation: The red spots indicate the position of laser focus. (a) untreated;(b) treated; (c) laser beam is shut off.

If ALS-CNC driven by focused laser beam can be controlled, this kind of contiguous self-oscillation can be used for local liquids stirrer at the nanometer scale in chemistry and medical science. In addition, CNCs have shown excellent electrical conductivity properties. Hayashida et al. shows that the electrical conductivity of CNCs is about100S/cm [35]. Hence, achieving a CNC switch controlled by laser irradiation in nanodevices or nano/micro electro-mechanic systems is possible.

4. Conclusion

We have discovered that the mechanical oscillation of a long single carbon nanocoil immersed thoroughly in water can be driven by focused laser beam. The mechanism of oscillation is the combining effect of trapping force from laser irradiation, the restores force from CNC flexibility and the damping force from liquid. In addition, we have found that self-oscillation does not occur by laser irradiation on ALS-CNC in air, and the CNC deflects only under prolonged laser irradiation. Our results clearly show that ALS-CNC in air can be used as an electricity switch controlled by laser irradiation in micro-systems or MEMS. Our work provides a new idea for the optically driven movement of CNCs.

Acknowledgments

This work is supported by Natural Science Foundation of Liaoning Province (grant No. 20180540081). This work is supported by the Seed Funding from Scientific and Technical Innovation Council of Shenzhen Government (grant No.2017010), the Open Fund of the Guangdong Provincial Key Laboratory of Optical Fibre Sensing and Communications (Jinan University), Natural Science Foundation of SZU (grant No.2017010).

References

[1] Kejie F, Jie L, Anja M, Matthew M, Florian M, Aashish C, Oskar P 2017 Nature Pysics 10 465
[2] W. D. P, 1998 Rev. Mod. Phys, 70 721
[3] Eriksson E, Sott K, Lundqvist F, Sveningsson M, Scrimgeour J, Hanstorp D, Goksör M, Granéli A 2010 Lab Chip 10, 617
[4] M. C. Z, X. B. W, J. H. Z, Z. Q. W, and Y.M. L 2013 Nat. Commun 4 1768
[5] K. W, E. S, P. S and K. B. C, 2011Nat. Commun 469
[6] L. Y, Y. Y. H, X. Z. Z and L. T 2017 Optics Express 25 18907
[7] Yulong L, Ribbing W, Jing Z, Ahin Kaya ¨Ozdemir S, Y. Lan,Franco N, and Yuxi L 2017 Phys. Rev. A 95 013843
[8] Liugang S, Hao X, Suhail Z and Ying W 2017 Phys. Rev. A 950 33803
[9] Yanliang H, Ying L, Yao C, Xiaoke Z, Junmin L, Jiangnan X, Shuqing C, Dianyuan F, Optik 2018 175 148
[10] Wurm G and Krauss O 2006 Phys. Rev. Lett.96, 134301
[11] Shvedov V, Davoyan A. R, Hnatovsky C, Engheta N, Krolikowski W 2014 Nat. Photonics 8 846
[12] Lin J, Hart A. G, Li Y.Q 2015 Appl. Phys. Lett. 106 171906
[13] Ashkin A 1970 Phys. Rev. Lett. 24, 156(1970)
[14] Ashkin A and Dziedzic J.M 1971 Appl. Phys. Lett., 19 283
[15] Degen C. L, Reinhard F and Cappellaro P, Reviews of Modern Physics 2017 89 035002
[16] Xu H, Kemiktarak U, J Fan, Ragole S, Lawall J, Taylor J.M 2017 Nature Communications 8 14481
[17] Jiagui W, ShuWei H, Yongjun H, Hao Z, Jinghui Y, Jiaming L, Mingbin Y, Guoqiang L, Dim lee K, Shukai D, Chee wei W 2017 Nature Communications 8 15570
[18] Marcelo W, Nathanael L, W. Y, Tayyaba F, Fatemeh F. S, Joseph E. L, Mark R. F, Paul Barclay E 2017 Nature Nanotechnology 12, 127
[19] Gilsantos E, Labousse M, Baker C, Goetschy A, Hease W, Gomez C, Lemaître A, Leo G, Ciuti C, Favero I 2017 Phys. Rev. Lett. 118 063605
[20] Jinsheng L, Hangbo Y, Lina Z, Yuanqing Y, Si L, Qiang L, Min Q 2017 Phys. Rev. Lett. 118 043601
[21] Lee S.H, Roichman Y and Grier D.G 2010 Opt. Express. 18 6988
[22] Shalin A.S, Sukhov S.V, Bogdanov A.A, Belov P.A and Ginzburg P 2015 Phys. Rev. A. 91 063830
[23] Bogdanov A.A, Shalin A.S, Ginzburg P 2015 Sci. Rep. 5 15846

[24] Lin J, Hart A.G and Li Y.Q 2015 Appl. Phys. Lett. 106, 171906
[25] Kajorndejnukul V, W.Q. Ding, S. Sukhov, C.W. Qiu, A. Dogariu, Nat. Photonics. 7, 787 (2013)
[26] D.B. Ruffner, D.G. Grier, Phys. Rev. Lett. 109, 163903 (2012)
[27] S.Yanming, P. Lujun, L. Yuli, S. Tao, Applied Surface Science 345, 428(2015)
[28] S.Yanming, L. Yuli, P. Lujun, Advanced Materials Research 818, 141(2013)
[29] Jiangtao L, Lujun P, He M and Dawei L 2010 Proc. of SPIE. 7659 765905
[30] Mani Rajan M.S and Bhuvaneshwari B.V 2018 Optik 175 39
[31] Ge J. H 2006 Journal of Clinical and Experimental Medicine 2 146
[32] Chen X, Dikin S. A, Ding W.R, Ruoff S, Pan L, Nakayama Y 2003 Nano. Letters 3, 1299
[33] Dawei L, Lujun P, Juanjuan Q, He M 2009 Adv. Mater 60 251
[34] Qin Z, Lujun P, He M 2012 Diamond and Related Materials 22 33
[35] Hayashida T, Lujun P, and Nakayama Y 2002 Physica B Physics of Condensed Matter 323 352

Research status of on-line monitoring of laser metal deposition

Zhaojun Jiang, Jun Wang*

Wuxi Institute of Technology, Wuxi, 214121, R.P. China

boher@126.com

Abstract. The process of laser metal deposition was easy to be affected by fluctuations of environment and process parameters, leading to the deviation of melt pool temperature or deposition morphology and so on. Accumulation of small deviation would result in the instability of deposition process and lead to the failure of forming. Therefore, the on-line monitoring of laser metal deposition process was studied, involving thermal history of forming parts, temperature and size of melt pool, deposition morphology and forming defects, and so on. In this paper, the on-line monitoring technologies, such as thermocouple temperature measurement, infrared thermal imaging, high-speed camera and other monitoring methods, were introduced, and the open-loop monitoring, closed-loop feedback control and closed-loop feedforward control technologies were illustrated in detail. According to the research situation, the future development of on-line monitoring in laser metal deposition was prospected.

Keywords: laser metal deposition, on-line monitoring, infared temperature measurement, thermal imaging, high-speed photography

With the development of science and technology, new requirements such as high efficiency, high performance and green energy saving have been put forward for modern manufacturing[1]. Traditional manufacturing technology has gradually been unable to meet the current new needs. Laser metal deposition technology (LMD), which is also called laser cladding, laser rapid prototyping, provides a new way for modern industrial production. It has great application value in the manufacture of high-complex parts[2], the manufacture and repair of large load-bearing structural parts[3] and the coating modification of material surface[4]. It can not only shorten the construction period, reduce costs, save strategic precious metal materials, but also protect the environment. And the performance of the parts can exceed the level of the traditional parts.

LMD forming process is a micro-pool establishment and non-stationary solidification process. The temperature, shape and size of the molten pool are easily affected by various factors. The accumulation of fluctuations may cause instability of deposition process and ultimately lead to failure of forming. However, these interference factors are extremely complex in the actual environment, and it is difficult to completely eliminate them. Based on this, researchers have developed a variety of monitoring technologies in order to maintain the stability of the forming process through closed-loop control of the temperature, shape and size of the molten pool.

There are five main monitoring techniques in LMD: (1) temperature measurement by thermocouple, (2) temperature measurement of molten pool by infrared thermometer, (3) temperature field measurement of molten pool by thermal imager, (4) molten pool melting transition behavior monitoring by high-speed camera, (5) Other monitoring methods such as laser scanning. This paper will introduce the research status of monitoring technology in LMD from these aspects.

1. Monitoring Technology

1.1. Thermocouple Temperature Measurement Technology

Thermocouple is the most traditional temperature probe, which is accurate and cheap. The temperature measurement method is single point contact temperature measurement, which requires direct contact with the material, causing it unable to be used for temperature measurement of molten pool. In the LMD process, thermocouples are usually used to measure the thermal history of printed parts, and the data are used for finite element simulation or analysis of the structure transformation of printed parts.

Ya et al.[5] simulated the laser cladding process by finite element method, and monitored the thermal history at different positions of the substrate using thermocouples, as shown in figure 1 (a). Finally, the error between simulation results and experimental measurements is less than 5%, which verifies the accuracy of the simulation calculation, as shown in figure 1 (b).

Heigel[6] and Ma[7] also used thermocouples to measure the thermal history at different positions of the substrate during wire-feed and powder-feed cladding, and compared it with the simulation results to analyze the energy absorption during laser cladding. Khan[8] and Meng[9] analyzed the microstructural evolution of materials during deposition using the thermal history measured by thermocouples.

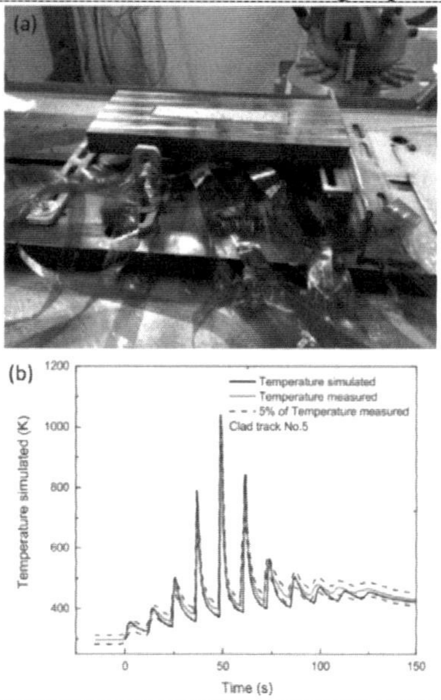

Figure 1. (a) Thermocouple temperature measuring device; (b) thermal history from thermocouple measurement and simulation[5]

1.2. Infrared Temperature Measurement Technology

Infrared thermometer detects temperature by infrared signal emitted by heating object, which is non-contact. Infrared thermometer is mainly divided into two kinds: one is monochromatic thermometer, which receives only one bands of infrared light and determines the temperature of the object by the intensity of the light; the other is colorimetric thermometer, which receives two bands of infrared lights and determines the temperature by the ratio of the strength of the two bands. Because infrared light is vulnerable to environmental factors such as smoke and dust, monochromatic thermometer is vulnerable to interference. Colorimetric thermometer has strong anti-interference because it takes the ratio of the strength of two bands.

Infrared thermometer, different from thermocouple, is a non-contact single-point temperature sensor, so it is often used to detect the temperature of molten pool in the LMD process. Song et al.[10] integrated the infrared thermometer into the printing system to keep it moving synchronously with the cladding head and the molten pool, so as to monitor the temperature change of the molten pool and realize the closed-loop control of laser power accordingly.

Infrared thermometer can also replace thermocouple to monitor the thermal history of printed parts. As shown in figure 2, Muvvala et al.[11,12] used infrared thermometer to collect the fixed-point thermal history and analyzed the microstructure evolution during cladding. Infrared thermometer can collect not only the temperature below the melting point, but also the temperature within a certain range above the melting point, which is more advantageous than thermocouple temperature measurement.

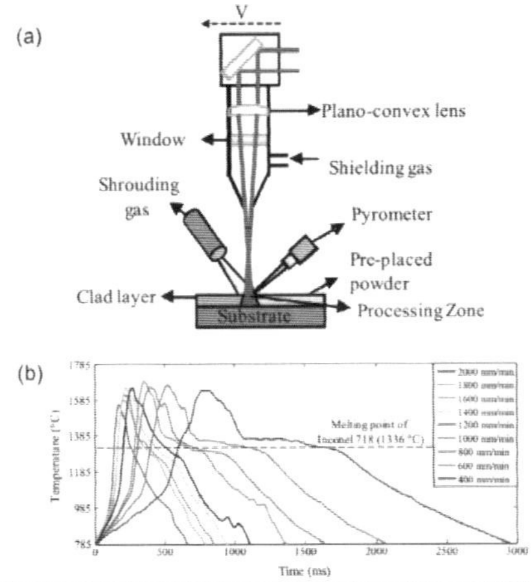

Figure 2. (a) Schematic diagram of temperature measurement principle of infrared thermometer; (b) thermal history at different scanning speeds[11]

1.3. Thermal Imaging Technology of Temperature Field

The signal intensity of infrared or visible light found at different temperature points of the object under test is different. When imaging in charge coupled device (CCD) or complementary metal-oxide-semiconductor (CMOS), each point presents different gray or brightness equivalents. According to the signal intensity of each point in the image, the temperature field of the object under test can be analyzed.

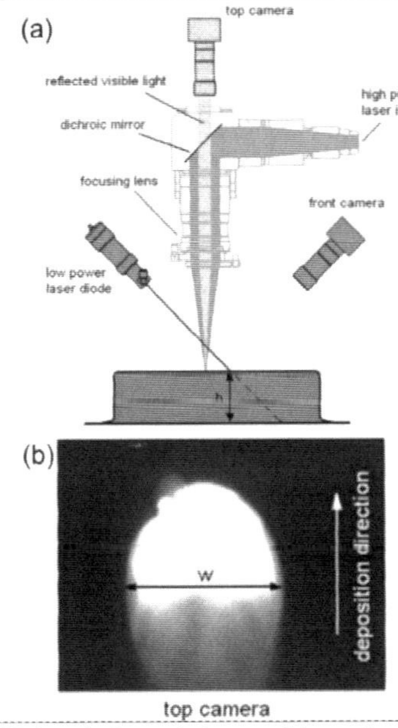

Figure 3. (a) Schematic diagram of CCD thermal imaging device; (b) schematic diagram of molten pool image and width monitoring[13]

Using infrared thermal imaging, not only the temperature, but also many other information, such as the shape and size of the molten pool, can be obtained by image analysis. As shown in figure 3, Heralić et al.[13] designed a CCD thermal imaging system to monitor the temperature field of the molten pool in wire laser additive manufacturing. The width of molten pool was calculated by using the image of molten pool temperature field, and the closed-loop control of laser power was carried out, which maintained the stability of the forming process. Yong et al.[14] monitored the temperature field of molten pool in wire-feed cladding by CCD, and extracted the area, width and length of molten pool by calculation. Lei[15] and García de la Yedra[16] carried out similar studies. In addition, Govekar[17] and Kuznetsov[18] used infrared thermal imaging to observe the droplet formation process at the tip of wire

during wire-feed rapid prototyping, which provided a reference for the adjustment of process parameters.
In addition to analyzing the shape and size of molten pool, some researchers used infrared thermal imaging technology to monitor forming defects. Ding et al.[19] studied the thermal behavior of wire laser additive manufacturing by using infrared thermal imager, and predicted the deposit width and poor fusion defects during deposition. Khanzadeh et al.[20] proposed a dual statistical process monitoring method based on infrared thermal imager, which combines image feature extraction and dimension reduction to identify local defects or anomalies in deposition process.

1.4. High-speed Photography Technology

High-speed camera has very high frame rate. With special filter lens and supplementary lighting equipment, it can take instantaneous phenomena such as melting, transition, solidification and splash of materials in the process of LMD, and then playback them slowly, which can provide the most intuitive visual image data for research.
Liu et al.[21] built a high-speed monitoring system in hot wire laser deposition, and observed the melting transition process of wire and the influence of different hot wire voltage on arc striking, as shown in figure 4. Torkamany et al.[22] used high-speed camera to observe the melting transition process from wire to keyhole, studied the influence of different process parameters on the shape of wire tip and metal melting transfer, and revealed the interaction mechanism between transverse oscillation laser and metal. Zhang[23] used high-speed camera to capture the image of molten pool, and compared the shape of molten pool with that calculated by finite element method.

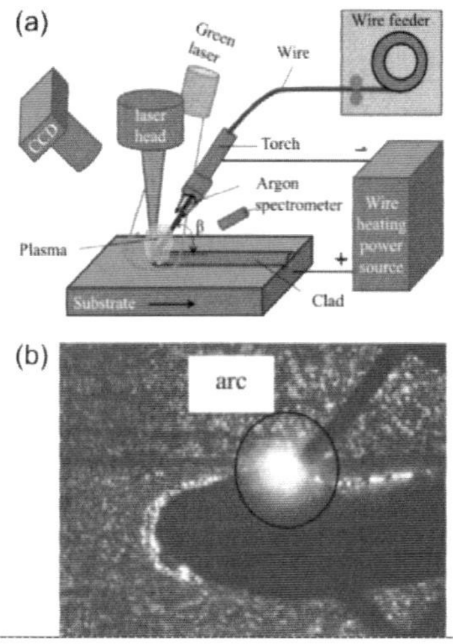

Figure 4. (a) Schematic diagram of high-speed photography; (b) real-time image of molten pool under hot wire condition[21]

1.5. Other Monitoring Techniques

In addition to the above types of monitoring methods, there are some special monitoring methods in LMD. Heralić et al.[24] designed a system for measuring the surface topography of the previous deposited layer by using diode laser. Based on this system, the wire-feed speed was controlled by iterative control algorithm, and then the deposition height of the next layer was controlled to ensure the stability of the deposition process. As shown in figure 5, Hagqvist et al.[25] used the resistivity of liquid bridge to monitor the current surface topography in real-time in wire-feed LMD. Through closed-loop control system, the surface morphology was smoother. Heigel et al.[26] used laser distance measuring instrument to monitor the substrate deformation in real time during cladding process.

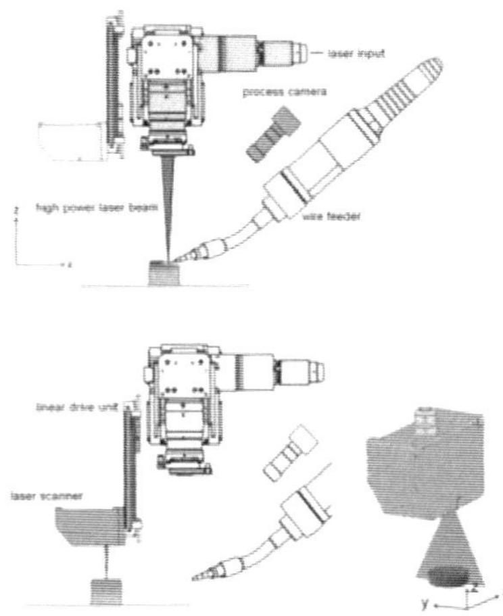

Figure 5. Schematic diagram of laser scanning detection for surface morphology[24]

1.6. Compound Monitoring Technology

Many of these monitoring methods are not used alone. Heigel et al.[6] simultaneously used thermocouples and laser ranging sensors to monitor the thermal history and deformation of the substrate. Song et al.[10] used both CCD and infrared thermometer to control the deposition height and the temperature of molten pool in a closed-loop way. Heraličet al.[13] also used thermal imaging technology of temperature field and laser diode profile temperature measurement technology to control the width and height of molten pool. The composite application of different monitoring technologies increases the complexity of the monitoring system, but also provides more abundant information, which is the trend of future development.

2. Application Mode

On-line monitoring is an important assistant method of LMD. Its main purpose is to improve the forming stability and quality. However, the application mode of monitoring technology is diverse. At present, there are three main applications of on-line monitoring: (1) open-loop monitoring, (2) closed-loop feedback control, (3) closed-loop feedforward control. Feedback control and feedforward control belong to closed-loop control technology, but they are different in detail. Therefore, they are divided into two sub categories for detailed introduction.

2.1. Open-loop Monitoring Technology

Open-loop monitoring is not a process control method, but just a data collection method. Only after manual analysis, the measured data can be adjusted to improve the forming stability and quality. For example, Liu et al.[21] directly observed the phenomenon of arc starting at the tip of metal wire under different hot-wire voltages by using high-speed camera, so that the limit value of hot-wire voltage was directly obtained, so as to achieve the purpose of adjusting process parameters. Some technologies, due to their own limitations, are only used to study the mechanism of microstructure evolution. For example, Ya[5], Heigel[6] and Ma[7] used thermocouple to measure thermal history, mainly to compare with the finite element simulation calculation result or to analyze the microstructure evolution, but it could not provide an effective reference for process parameter adjustment. This is because the thermocouple can only monitor the thermal history of a certain point in the non-molten pool and cannot reflect the actual state of the molten pool.

Open-loop monitoring can provide a lot of data for process optimization, but it is inefficient because it requires manual analysis. Moreover, the open-loop monitoring data cannot reflect the real-time state of the forming process. Only an overall adjustment of the process parameters can be made by open-loop monitoring, and the process parameters in a certain stage or instantaneous process parameters cannot be finely adjusted, which cannot effectively improve the anti-interference ability of the forming process.

2.2. Closed-loop Feedback Control Technology

The closed-loop feedback control is a kind of monitoring technology which adjusts the process parameters online in real time after the monitoring data is analyzed and fed back to the numerical control system. As shown in figure 5, Song et al.[10] analyzed the temperature and height of the molten pool measured by pyrometer and CCD respectively, and adjusted the laser power in real time by using the feedback system, thus improving the stability of forming process. Through CCD thermal imaging, Heralić et al.[13] analyzed the width of molten pool and controlled the laser power in real time online.

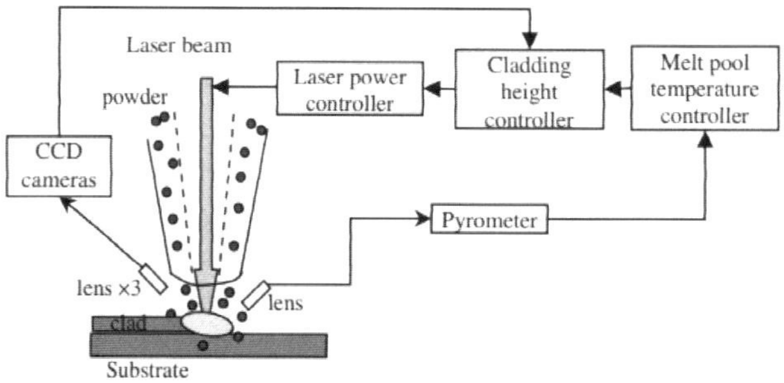

Figure 6. Closed-loop feedback control system based on infrared temperature measurement and thermal imaging[10]

The closed-loop feedback control is a promising monitoring method, which can significantly improve the forming stability and quality. Moreover, the addition of the feedback control system does not affect the forming efficiency. The closed-loop feedback control is suitable for controlling process parameters with fast response speed. As shown in figure 7, when the laser power is increased instantly, its influence is quickly reflected on the temperature and size of the molten pool, with no or little lag effect[13].

2.3. Closed-loop Feedforward Control Technology

The closed-loop feedforward control is a monitoring technique that monitors and analyzes and adjusts process parameters before the end of each layer or section of deposition. This monitoring technique is usually used to control process parameters with hysteresis effect. As shown in figure 7, the wire-feed speed will be reflected in the deposition height in the time of about 3 s after it is increased, with a certain lag effect[13].

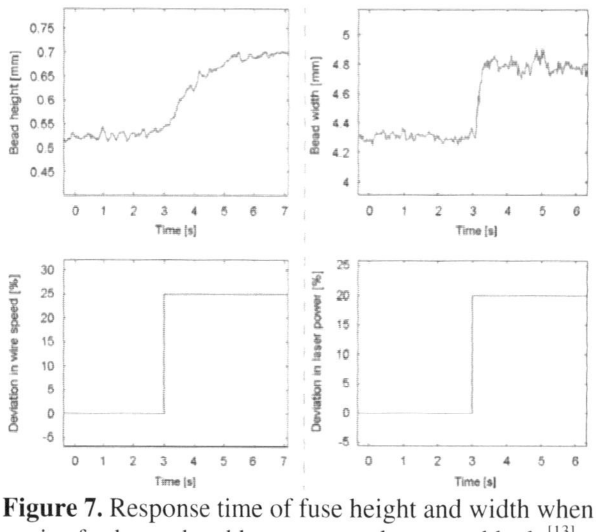

Figure 7. Response time of fuse height and width when wire-feed speed and laser power change suddenly[13]

Therefore, feedforward technique is adopted by Heralić for monitoring. The height of the previous layer is measured before each section is deposited, and the wire-feed speed is adjusted in advance. As shown in figure 5, Heralić [24] also use a similar feedforward technique in the monitoring of surface morphology. Before each layer is deposited, the surface morphology of the previous layer is measured first, and then process parameters of each segment of the path of the next layer are adjusted. The closed-loop feedforward control has a good effect on the control of process parameters with hysteresis, which is complementary to the feedback control. However, because part of feedforward control needs to stop the deposition process first before the process parameters is adjusted, it results in the reduction of forming efficiency.

3. Conclusions And Prospects

In recent years, on-line monitoring technology has made great progress, which provides powerful data support for LMD research and plays a very important role in its research. In these monitoring methods, thermocouple temperature measurement, infrared temperature measurement and temperature field imaging are temperature-related detection. Thermocouple temperature measurement has the greatest limitation. It can only measure the temperature of fixed point, and can not get more information. Infrared thermometer can completely replace the work of thermocouple to monitor the thermal history. At the same time, it can also track the temperature of molten pool in real time. Temperature field imaging can get more information, so it is also the most widely used. High-speed cameras are mainly used to observe the melting transition process of materials. The information obtained is relatively small, the shooting time is relatively short, and the equipment is extremely expensive. Therefore, their applications are relatively small. Among other monitoring methods, laser scanning is a more effective feedforward control technology and a more promising monitoring technology. These monitoring technologies have their own focus. Current research is no longer limited to the use of a single monitoring technology and simple data monitoring, but more and more inclined to the integration of multi-technology and closed-loop control.

At present, the research of LMD online monitoring is still weak in closed-loop control. First of all, most of the current rapid prototyping technology is still not mature enough, and still stays at the research stage, such as wire laser additive manufacturing, which has not yet reached the stage of industrial application, and the demand for closed-loop control is still small. Secondly, there is no dedicated independent control module to ensure its efficient data acquisition, analysis and feedback. Therefore, this paper holds that the on-line monitoring technology will make great progress in the following aspects: (1) the development and application of high-speed analysis module or software for on-line monitoring data to solve the shortcomings of the current monitoring technology, such as slow data analysis and lagging response; (2) the development and application of new monitoring algorithms such as in-depth learning to realize intelligent control of forming process; (3) the development and application of high-precision and high-sensitivity monitoring sensors can provide more accurate data and faster response for LMD process.

References

[1] Zhao Y, Liu C, Guo Y, et al 2018 *J. Progress in Natural Science: Materials International* **28** 483–488.
[2] Xiao Z, Yang Y, Xiao R, et al 2018 *J. Materials & Design* **143** 27–37.
[3] Kottman M, Zhang S, McGuffin-Cawley J, et al *2015 J. JOM - Journal of the Minerals, Metals and Materials Society* **67** 622–628.
[4] Abioye T E, McCartney D G and Clare A T 2015 *J. Journal of Materials Processing Technology* **217** 232–240.
[5] Ya W, Pathiraj B and Liu S 2016 *J. Journal of Materials Processing Technology* **230** 217–232.
[6] Heigel J C, Gouge M F, Michaleris P, et al 2016 *J. Journal of Materials Processing Technology* **231** 357–365.
[7] Ma P, Wu Y, Zhang P, et al 2019 *J. The International Journal of Advanced Manufacturing Technology* 1-13.

[8] Khan M, Maurya K, Thawari N, et al 2018 *J. IOP Conference Series: Materials Science and Engineering* **455** 012129.
[9] Gouge M F, Heigel J C, Michaleris P, et al 2015 *J. The International Journal of Advanced Manufacturing Technology* **79** 307–320.
[10] Song L, Bagavath-Singh V, Dutta B, et al *2012 J. The International Journal of Advanced Manufacturing Technology* **58** 247–256.
[11] Muvvala G, Patra Karmakar D and Nath A K 2017 *J. Optics and Lasers in Engineering* **88** 139–152.
[12] Muvvala G, Patra Karmakar D and Nath A K 2018 *J. Journal of Alloys and Compounds* **740** 545–558.
[13] Heralić A, Christiansson A-K, Ottosson M, et al 2010 *J. Optics and Lasers in Engineering* **48** 478–485.
[14] Yong Y, Fu W, Deng Q, et al 2017 *J. Journal of Manufacturing Processes* **28** 364–372.
[15] Lei K, Qin X, Liu H, et al 2018 *J. Optics and Lasers in Engineering* **110** 89–99.
[16] García de la Yedra A, Pfleger M, Aramendi B, et al 2019 *J. Structural Control and Health Monitorin* **26** 2291.
[17] Govekar E, Bezgovšek J and Grabec I 2011 *J. Physics Procedia* **12** 421–427.
[18] Kuznetsov A, Jeromen A and Govekar E 2014 *J. CIRP Annals* **63** 225–228.
[19] Ding X P, Li H M, Zhu J Q, et al 2017 *J. Infrared Physics & Technology* **81** 166–169.
[20] Khanzadeh M, Bian L, Shamsaei N, et al 2016 *Proc. Int. Conf. on 26th Annual International Solid Freeform Fabrication Symposium* p 1487-1494.
[21] Liu S, Liu W, Harooni M, et al 2014 *J. Optics & Laser Technology* **62** 124–134.
[22] Torkamany M J, Kaplan A F H, Ghaini F M, et al 2015 *J. Optics & Laser Technology* **69** 104–112.
[23] Zhang Z, Farahmand P and Kovacevic R 2016 *J. Materials & Design* **109** 686–699.
[24] Heralić A, Christiansson A-K and Lennartson B 2012 *J. Optics and Lasers in Engineering* **50** 1230–1241.
[25] Hagqvist P, Heralić A, Christiansson A-K, et al 2014 *J. Optics and Lasers in Engineering* **54** 62–67.
[26] Heigel J C, Michaleris P and Palmer T A 2015 *J. Journal of Materials Processing Technology* **220** 135–145.

IOP Conf. Series: Materials Science and Engineering **605** (2019) 012021 doi:10.1088/1757-899X/605/1/012021

Creep-rupture behavior of a nickel-based single crystal superalloy

Weiwei Liu[1,*], Yuanyuan Guo[1], Mai Zhang[1,2], Jichun Xiong[1], Jian Zhang[1]

1. Science and Technology on Advanced High Temperature Structural Materials Laboratory, Beijing Institute of Aeronautical Materials, Beijing 100095, P.R. China
2. School of Materials Science and Engineering at University of Science and Technology Beijing 100083, P.R. China)

E-mail: liuww_80@sina.com

Abstract. A first generation single crystal superalloy was employed to research the high temperature creep behavior. Scanning electron microscope (SEM), transmission electron microscope (TEM) are used. The main results are summarized below: There are two mechanisms for the steady creep. The first one is superlattice dislocation cutting γ' precipitates, which is the main mechanism for 1000℃ creep behavior. In this case, the creep rate increases when the γ/γ' microstructure undergoes topological inversion, which leads to a rapid increase in the density of γ' cutting dislocation. The second mechanism is transverse climb of a/2<110> interfacial dislocation, which is the main mechanism for 1100℃ creep behavior. In this case, climb of interfacial dislocations produced most of the steady creep strain. A by-product is the vacancies diffusing along the interfaces and low angle boundary to pores.

Key words: single crystal ;superalloy ;high temperature creep ;dislocation ;porosity

Nickel-based single crystal superalloys have been widely used in the manufacturing of aero-engine turbine blades because of its excellent high temperature creep strength and fatigue strength, as well as good thermal corrosion resistance and oxidation resistance[1,2]. With the increasing temperature of the advanced aero-engine turbine, high temperature creep deformation and fracture become one of the main failure forms of single crystal turbine blades. The creep behavior of nickel-based single crystal superalloy under typical temperature stress has been widely studied [3-8], and the research mainly focuses on the microstructure

evolution, the deformation mechanism of different creep stages and the prediction model of creep life, in which the directional coarsening of γ′ phase is a typical phenomenon in the medium and high temperature creep process of single crystal alloy. However, there is no unified conclusion about its effect on the properties of alloys. Furthermore, the creep deformation mechanism of single crystal alloy at about 760 °C has been thoroughly studied[9,10]. The research indicates that the creep behavior and deformation mechanism of single crystal superalloys are influenced by experimental temperature, stress range, loading mode and so on. This is mainly caused by the interaction between dislocations and γ′ particles. With the changes of these factors, both of the creep and deformation mechanism of the alloy are changing[11-19]. Generally, there are still few researches on creep deformation mechanism of single crystal alloys at high temperature level, i.e. 1100 °C, especially existing plenty of problems to be studied in the third stage of creep deformation and creep fracture mechanism.
In this present work, a nickel-based single crystal superalloy was employed to study the high temperature creep under low applied stress behavior at 1000 °C and 1100 °C. The high temperature creep behavior, creep deformation, fracture characteristics and mechanism of single crystal alloy were systematically studied and analyzed. The different of creep mechanisms at two temperature were also compared. Results documented in this study should be able to provide an experimental basis for the utilization of single crystal alloy under high temperature conditions, and further improve the creep theory of nickel-based single crystal superalloy.

1. Experimental materials and procedures

The experimental material was the first generation of nickel-based single crystal superalloy with chemical compositions of Ni-Cr-W-Ta-Al-Co-Ti-Mo system. The constant load tensile creep test was carried out on high temperature creep machine with applied stress ranging from 120 to 220 MPa, and the experimental temperature were 1000 °C and 1100 °C. The furnace temperature was controlled by the upper, middle and lower platinum-rhodium thermocouples which placed in the normal distance part of the specimens, and the temperature controlled within ±2 °C. creep strain was measured by an extensometer connected to the shoulders of creep specimens, the accuracy was 5×10-4~5×10-5and experimental environment was air. Most of specimens were pulled to fracture, part of specimens was subjected to creep interruption test for the observation of microstructure.
The specimens after creep fracture and interruption test were observed and analyzed by JSM-6301 field emission scanning electron microscopy and Tecnai G 20 transmission electron microscopy (TEM). When preparing TEM specimens, 0.5 mm thick sheets were cut along the direction perpendicular to the tensile axis and 5-10 mm away from the tensile fracture. Firstly, mechanically ground to below 50 μm, then double jet electrolysis was used to reduce the thickness. The electrolyte was composed of 10% perchloric acid and 90% alcohol.

2. Results and discussion

2.1. High temperature creep properties of experimental alloy

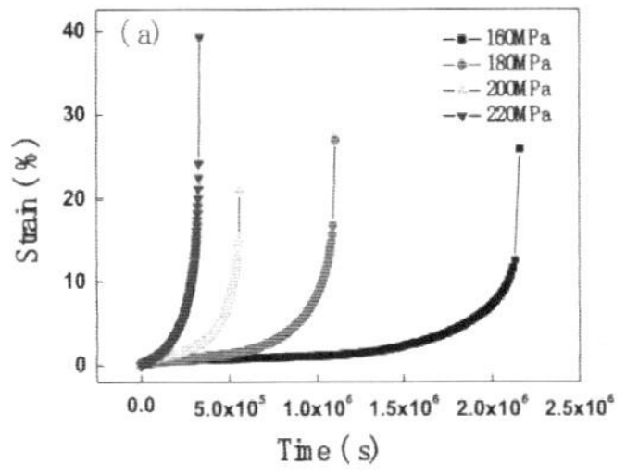

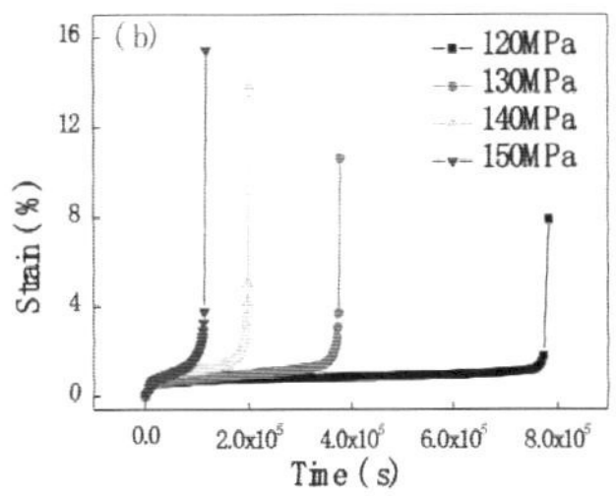

Figure 1. Typical creep curves of experimental alloy:
(a) 1000 °C and (b) 1100 °C

Figure 1 shows the typical creep curves of experimental alloy under different stress levels at 1000 °C and 1100 °C. As can be seen from the strain vs. time relationship, the main characteristics of high temperature creep behavior of experimental alloy are as follows:

(1)Both the creep curves at 1000 °C and 1100 °C exhibit three stage of creep characteristics, i.e. initial creep, steady creep and accelerated creep. The duration of creep steady-state stage at 1100 °C accounts for a higher proportion of the whole creep life, and the creep curve shows a clear platform shape.

(2)The creep extension decreases obviously when the experimental temperature increasing from 1000 °C to 1100 °C. Under the conditions of 1000 °C and 160 MPa – 220 MPa, the fracture extension ranges from 20% to 40%. Compared with the former, the extension at 1100 °C ranges from 8% to 16%, which is obviously smaller than that of the former.

(3)In the creep test at 1000 °C, the strain rate rises slowly after alloy entering the third stage of creep. While in the test at 1100 °C, the strain rate rises rapidly after the beginning of the third stage of creep, and fracture occurs in a short period of time.

2.2. Microstructural evolution of creep process

As can be seen from figure 2a and b, the shaped raft of directional coarsening of γ' phaseoccurs during the creep test at high temperature ranging from 1000 °C to 1100 °C. The alloy belongs to negative misfit alloys, hence the raft structure perpendicular to the direction of applied stress formed under tensile stress. The dislocation configuration observed by TEM shows that complete and regular dislocation network at the γ/γ' interface is rapidly formed during creep at 1000 °C and 1100 °C. The formation of dislocation network originates from the reaction between dislocations at the γ/γ' interface, which generally acts as impediment for dislocation movement and reduces creep rate. figure 2c and d reveal that superlattice dislocation cutting γ' phase occurs at both temperature in the post stage of creep, which is the main deformation mechanism of creep steady-stare stage. In contrast, the greater number of dislocations are cut into γ' phase and show multiple morphologies under 1000 °C and higher stress condition. However, there are only few dislocations are cut under 1100 °C and higher stress condition, and mainly exist in a single form of screw dislocations. This is the biggest

IOP Conf. Series: Materials Science and Engineering **605** (2019) 012021 doi:10.1088/1757-899X/605/1/012021

difference between the creep deformation mechanism at 1100°C and 1000 °C. It indicates that large number of dislocations can not be cut into γ' phase due to insufficient applied stress, hence the dislocation configuration does not change significantly during the whole creep process.

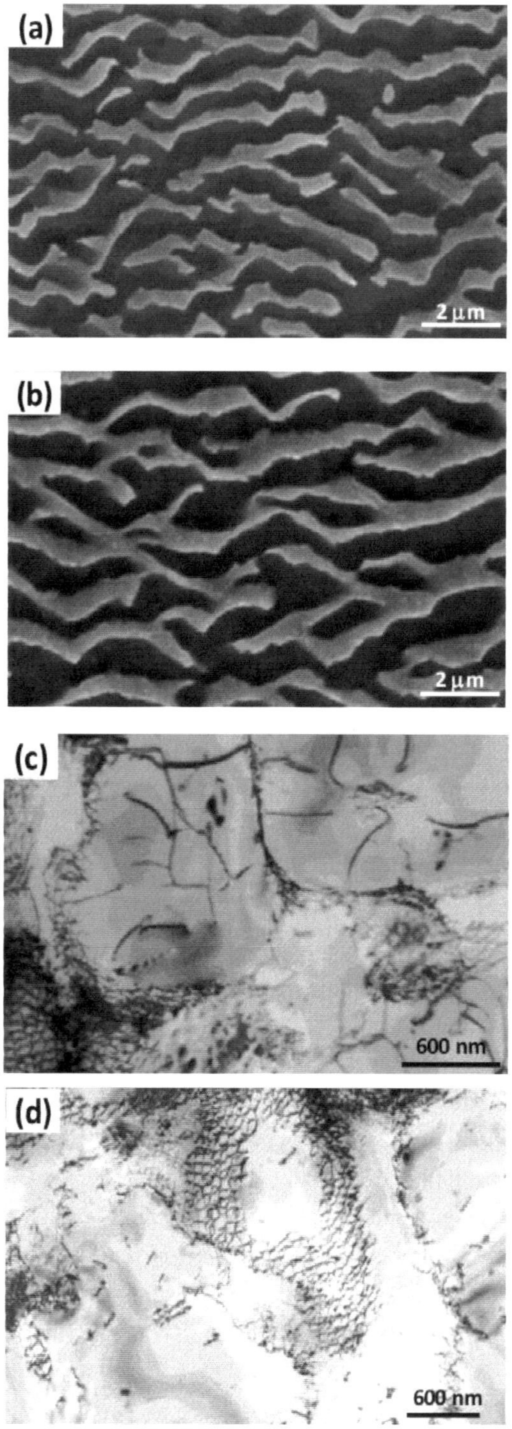

Figure 2. Microstructural evolution after high temperature creep fracture

2.3. Behavior of pore and creep fracture

It can be seen from the fracture morphology of experimental alloy after creep fracture that fracture modes at 1000 °C and 1100 °C are the same as microporous aggregate fracture (figure 3). The creep fracture of is composed of abundant square facets, the orientation of planar edge is <110> direction, and the small circular hole at the center of plane is the origin of the crack. As the temperature rises to 1100°C, the contour of teas plane in the fracture gradually becomes blurry, dimple shape becomes irregular and the number of crack initiation points increases.

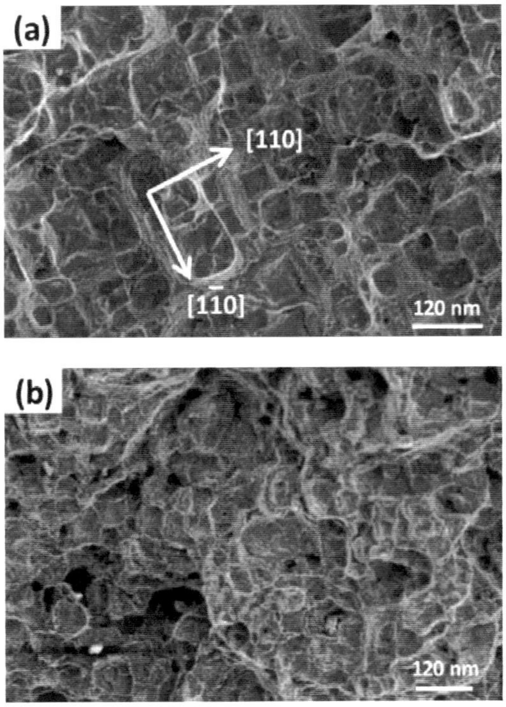

Figure 3. Fracture morphology of alloy after creep fracture

2.4. Analysis and discussion

In the temperature and stress ranges of 1000 °C, 160-220 MPa and 1100 °C, 120-150 MPa, the alloys exhibit typical deformation characteristics of the above series of single crystal alloys under high temperature and low stress creep, but it is worth noting that there are also some obvious differences in the creep behavior between the two experimental temperatures. The main manifestations are smaller proportion in the third stage of creep at 1100 °C, faster the deformation rate, and larger density of pores after creep fracture. However, the number of dislocations in the γ' phase and the creep elongation both are large in the third stage of creep at 1000 °C.

The controlling factor of creep in the initial stage is the slip of the a/2<110> dislocation loop in the {111} plane in the transverse matrix channel perpendicular to the applied stress [001] direction, due to the applied stress has the same direction as the γ/γ' dual-phase mismatch stress ondislocations. The widening of the transverse matrixchannel caused by the γ' phaseshaped raft will further promote the movements of the dislocation loop. The generation and operation of dislocations are affected by temperature. The number of dislocation sources at 1100 °C is much higher than 1000 °C, so the initial creep content is significantly higher than that of the latter. The results of initial creep are as follows: a large number of matrix dislocations deposit at the interface of dual-phase, begin to form dislocation networks, γ' phase transforms into a raftmicrostructure, and the creep rate decreases gradually under the influence of the dislocation networks.

As the resistance of matrix dislocation movements increases, the deformation rate of initial creep slows down and creep begins to enter steady state. During the creep steady state, raftmicrostructure completely hindersthe longitudinal matrix channel, making it impossible for the dislocations to continue moving in a way that bypass the γ' phase. An important recovery mechanism for maintaining deformation at this time is that dislocations enter the interior of γ' phase in pairs. As mentioned above, during the creep steady state, dislocations cut raft structure. The superlattice dislocations cutting γ' phase will annihilate with the opposite superlattice dislocations on the other side of raft structure, which is the main recovery mechanism of creep steady state.On the other hand, single dislocation cutting mechanism can not explain the formation and growth of pores during creep. The climbing of depositional dislocations at the γ/γ' dual -phase interface can be regarded as another recovery mechanism. Carry and Strudel[20] et al. firstly proposed that the γ/γ' interface dislocation climbing can be regarded as plastic deformation mechanism. Epishin and Link[21-22] et al. pointed out that after creep enters steady state, the 60° dislocations deposited at the γ/γ' interface can perform non-conservative movements along the direction perpendicular to the applied load, which is mainly dominated by the diffusion process. The non-conservative movements can be regarded as a combination of dislocation slipping and negative climbing, which leads to the alloy undergoing plastic deformation parallel to the applied load direction, and the negative climbing of dislocations causing a large number of vacancies to be discharged These vacancies are diffused to pores through the rapid channels such as the interface between raft and matrix, and low angle boundaries, resulting in the growth of pores. The increase of pore volume fraction after creep can be explained by this deformation mechanism.

In summary, there are two main deformation mechanisms after high temperature creep enters steady state, which are dislocations cutting γ' phase and interface dislocations climbing, respectively. Obviously, the dislocations cutting mechanism requires higher stress level, but the interface dislocations climbing mechanism is more dependent on the creep experiment temperature.

3. conclusions

In this work, a nickel-based single crystal alloy was employed to study the high temperature creep under low applied stress behavior at 1000°C, 160-220 MPa, 1100 °C, and 120-150 MPa. The main conclusions are as follows:

(1)The third stage of creep at 1000 °C lasts longer, deformation rate increases slowly, and the elongation of the alloy is larger. However, the deformation rate of the third stage of creep rapidly increases and fractures in a short time at 1100°C, and the total elongation decreases by about 50% compared with the former;

(2)There are two main deformation mechanisms in the high temperature creep steady state , which are dislocation cutting γ' phase and interface dislocation climbing, respectively. The influence of the two mechanisms on creep deformation varies according to temperature and stress conditions. The former requires larger applied stress to press dislocations cutting γ' phase, while the latter requires higher experimental temperature to promote diffusion;

(3)The main deformation mechanism of 1000°C and 160-220 MPa creep steady state is dislocations cutting γ′ phase. In the post stage of creep steady state, the γ/γ′ dual-phase structures become topologically inverted, and the continuous slipping of superlattice dislocation inside the γ′phase causes the aggravation of alloy deformation and creep entering the third stage;

(4)The main deformation mechanism of 1100 °C and 120-150 MPa creep steady state is the migration of γ/γ′ dual-phase interface dislocations. The dislocations slip and negatively climb perpendicular to the main stress axis, and discharge a large number of vacancies, and vacancies diffusing to pores result in the growth of casting pores and the initiation of new creep pores. The cracking of a large number of pores leads the deformation rate of the alloy to rise rapidly and fracture in the third stage of creep.

References

[1] Reed R C 2006 *M The superalloys: Fundamentals andapplications.*
[2] Tang Y, Huang M, Xiong J, et al 2017 *J. Acta Materialia* **126** 336-345.
[3] Ma S, Lv X, Zhang J, et al 2018 *J. Journa of Alloys and Compounds* **743** 372-376.
[4] Liu W, Liu S, Li J, et al 2014 *J. Materials Research Innovations* **18** 445-449.
[5] Wen Z X, Zhang D, Li S, et al 2017 *J. Journal of Alloys and Compounds* **692** 301-312.
[6] Shu D L, Tian S G, Liang S, et al 2010 *J. Journal of Materials Engineering* **45** 93-100.
[7] Li A N, Tian S G, Liang F S, et al 2010 *J. Journal of Materials Engineering* **1** 210-215.
[8] Wang M G, Tian S G, Yu H C, et al 2009 *J. Journal of Aeronautical Materials* **29** 98-102.
[9] Shollock B, Buffiere J, Curtis R, et al 2017 *J. Scripta Materialia* **36** 1471–1478.
[10] Rae C M F, Reed R C 2007 *J. Acta Materialia* **55** 1067-1081.
[11] Tian S, Ding X, Guo Z, et al 2014 *J. Materials Science and Engineering: A* **594** 7-16.
[12] Wollgramm P, D. Bl, Parsa A B, et al 2016 *J. High Temperature Technology* **33** 346-360.
[13] Huang M, Zhuo L, Xiong J, et al 2015 *J. Philosophical Magazine Letters* **95** 1-8.
[14] Pollock T M, Argon A S 1992 *J. Acta Metallurgica Et Materialia* **40** 1-30.
[15] Hafez h S M, Eggeler G, Raabe D 2013 *J. Acta Materialia* **61** 3709-3723.
[16] Probst-hein M, Dlouhy A, Eggeler G 1999 *J. Acta Materialia* **47** 2497-2510.
[17] Zhao Y S, Liu C G, Guo Y Y, et al 2018 *J. Progress in Natural Science:Materials International* **28** 483-488.
[18] Zhang J X, Wang J C, Harada H, et al 2005 *J. Acta Materialia* **53** 4623-4633.
[19] Huang M, Cheng Z, Xiong J, et al 2014 *J. Acta Materialia* **76** 294-305.
[20] Carry C, Strudel J L 1978 *J. Acta Metallurgica* **25** 767-777.
[21] Epishin A, Link T 2004 *J.Philosophical Magazine* **84** 1979-2000.
[22] Link T, Zabler S, Epishin A, et al 2006 *J. Materials Science & Engineering A* **425** 47-54.